Solar Cells Redefined The Promise of Quantum Dot Sensitization

Irwin Thomas

TABLE OF CONTENTS

2 SYNTHESIS AND CHARACTERIZATION OF TiO$_2$ NANOSTRUCTURES WITH CONTROLLED MORPHOLOGY FOR QUANTUM DOT

SENSITIZED SOLAR CELL APPLICATIONS 37

ABSTRACT

The demand for electric energy is rapidly increasing due to increasing world population, along with rising environmental issues resulting from burning of fossil fuels. Hence there is urging need to develop an alternate technology for generating electric energy using an ecofriendly and clean renewable sources. Photovoltaic (PV) technology is an efficient technology photoelectric effect which is based on converting light-to-electricity. Silicon dominates the PV industries since more than 85 % of the solar cells are made from Si. Highly expensive and complicated fabrication techniques deprived PV technology and led to development of simple, less expensive nanotechnology based novel solar cells like sensitized solar cells. Quantum dot sensitized solar cells are most prominent and attractive technology using semiconductor Quantum Dots (QDs) as sensitizers. Unique properties exhibited by QDs such as size dependent electronic structure, large extinction coefficients, and multiple exciton generation render them attractive candidates as photon harvesters.

A strategy to effectively harvest solar energy using QDSSC is to do a state of art research focussing on novel QDs and advanced photoelectrode design with proper band alignment of each layers of QDSSC. Recognition and development of QDSSC using new QD material and optimized cell components is a new platform to the researchers in order to develop an efficient solar cell. The main focus of the thesis is to design and develop a highly efficient & stable QDSSC by optimizing the components of the cell at relatively lower fabrication cost. The work mainly focusses on development of II-VI, III-V semiconductor QDs and to study its performance as sensitizers for QDSSC. Moreover, optimization of each components of QDSSC is highly imperative to improve the efficiency and overall performance of cell.

TiO_2 nanostructures are considered as a main material for photoanode of QDSSC. TiO_2 nanostructures with different morphologies were synthesized using sol gel and hydrothermal methods by varying growth period and temperature. Further, the experimental results of XRD, SEM, FTIR, RAMAN, XPS, optical and electrical studies demonstrated that the morphology of TiO_2 nanostructure has great influence on its optical and electrical properties.

Cadmium sulfide (CdS) QDs are widely used as sensitizer materials for QDSSC. The preparation of CdS QDs by greener approach was carried out mainly with an objective to control its toxicity and compared with CdS prepared by conventional SILAR method. For controlling toxicity, CdS QDs was prepared using Azadirachta Indica (A. indica) extract as an environment friendly and cost effective nontoxic stabilizing agent. The toxicity of CdS QDs was effectively suppressed as observed by hemolysis and cytotoxicity studies. As prepared green CdS was used as sensitizer for QDSSC and studied in comparison to that of SILAR prepared CdS based QDSSC. Both green CdS and SILAR CdS were loaded on to TiO_2 photoanode for QDSSC and its photovoltaic performance under AM 1.5G illumination was studied using solar simulator. Green CdS based QDSSC, exhibits high photocurrent density (10.61 mA/cm$_2$) compared to conventional cells (9.36 mA/cm$_2$).

Greater the light absorption of the sensitizer, more will be the photoconversion efficiency for QDSSC. One of the main reasons for low efficiency of QDSSC is the delimited performance of sensitizer materials which are active only in the visible regime of solar spectrum. While the solar light reaching earth, surface includes 8% of ultraviolet (UV) rays, 42% of visible light and 50 % of infrared (IR) light. Hence by effective utilization of IR region of solar spectrum using IR active QDs, could improve light absorption of sensitizers and therefore overall efficiency of QDSSC can be greatly enhanced. For improved light absorption with broadened solar spectrum coverage, Indium antimonide (InSb) QDs has been studied as a sensitizer of QDSSC for the first time. InSb QDs were successfully synthesized by optimizing the conditions using solvothermal method. The structural, morphological and optical properties of synthesized InSb QDs were investigated by XRD, SEM, TEM and UV-Vis-NIR spectroscopy. The fluorescence properties of InSb QDs were studied. The synthesized InSb QDs were used as sensitizer for QDSSC along with other cell components such as polysulphide – redox electrolyte, and platinum/CuS counter electrode (CE). Moreover, the effect of co-sensitization of InSb with visible active CdS QDs on the performance of QDSSC were analysed. Under AM 1.5G illumination, InSb/CdS co-sensitized QDSSC showed an enhanced photovoltaic characteristic with an efficiency of 4.94 %. The InSb QD layer broadens the light absorption range with reduced spectral overlap causing an improvement in light harvesting. Moreover, the co-sensitization of InSb with CdS effectively suppressed the surface defects with low recombination losses which resulted high efficiency.

Carbon chemistry, unfolds exciting opportunities to tune the optical and electronic properties of graphene material that can be effectively investigated for photovoltaics application. Graphene QDs (GQDs) were synthesised by a liquid exfoliation method and used as a passivating layer for CdS QDSSC. GQD passivated QDSSC showed relatively higher efficiency (4.06 %) compared to ZnS passivated QDSSC (3.23 %) with CuS CE. GQDs served as an effective passivating layer which suppressed charge recombination and as a bridge inhibiting back electron from electrolyte, leading to a significantly enhanced open circuit voltage (555.67 mV) compared to ZnS passivated QDSSC which resulted higher efficiency.

Further to improve QDSSC performance the polysulphide electrolyte was modified using Tetraethyl orthosilicate (TEOS) as additive with different vol. %. The J-V characteristics of CdS based QDSSC with optimum photoanode along with ZnS passivating layer were studied by adding different vol. % of TEOS into polysulphide electrolyte. QDSSC with 2.5 vol.% of TEOS have shown an improved efficiency of 1.6 % with Pt CE. On replacing Pt CE by CuS, the efficiency of QDSSC has improved significantly up to 5.5% due to more compatibility of polysulphide to CuS. Moreover, a comparative analysis of drop casted CuS and SILAR CuS CEs on the performance of QDSSC have been studied. The addition of TEOS effectively improved the passivation of QDs thereby suppressed the recombination of charge carriers which resulted high efficiency of QDSSCs.

CHAPTER 1 INTRODUCTION

INTRODUCTION

The tremendous rise in per capita energy consumption owing to the increasing world population and diminution of conventional energy sources resulted the huge demand for electric energy. The increasing global warming due to CO_2 emission from all over the world has stimulated alternate way for generating electricity instead of burning the fossil fuel. Hence there is a real need to develop alternate sustainable and renewable energy sources for present and next generation, which can solve both energy and environmental problems simultaneously. Major global sources for electric energy are oil (33.5 %), coal industries (26.8 %), natural gases (20.9 %), nuclear power plants (5.8 %) and other renewable energy sources (12.8 %). Recent researches suggest that by focussing on energy generation mainly from renewable energy sources, the global greenhouse gas emissions can be reduced upto 35 % by the year 2030, which leads to green environment (Jacobson *et al.* 2011). Therefore, it is highly essential to look for sustainable energy generation from alternate sources like hydrothermal, geothermal, biomass, wind energy and solar energies that are reliable and environment friendly (Bevrani *et al.* 2010).

Hydrothermal Energy

Main source of power is potential energy of water that is used to rotate the turbine for generating electricity. Potential energy is converted to kinetic energy and used to generate electric power. It is one of the green ways of generating the electricity as it does not emit any pollutant. However, it has certain limitations such as hydroelectric power stations need to be built at elevated area to allow water to fall from larger heights. Also, this power generation method depends on availability of water such as rainfall etc. Hydrothermal energy generation covers 2.4 % of total energy consumption in the world (Titirici *et al.* 2012).

Geothermal Energy

The main source of energy is from the core of earth crust which is at very high temperature. The thermal energy from interior core of earth is utilized as a thermal source for heating water on earth crust and for rotating turbine to generate kinetic energy and to electric power. Geothermal energy is source of renewable energy and it is independent on any weather condition and highly ecofriendly. It contributes 9 % of total world energy consumption Lee (2001).

Biomass Energy

Energy from chemical bonds of biowaste when extracted effectively by either chemical or biological methods releases energy in the form of CO_2, which is used to produce biomass by photosynthesis process. Organic materials prepared from plant constituents like aquatic vegetarian, organic waste, terrestrial are considered as biowaste. The biomass is a renewable source of energy that can be converted in to three forms of products such as electrical energy, chemical feed stock and even transport fuel. Biomass covers 10 % of global energy contribution. However organic components from animal waste can cause bacterial infections and diseases. Moreover, the cost required for processing waste and its management is quite expensive along with its requirement of large land area limiting its wide spread development (McKendry *et al.* 2002).

Wind Energy

Wind is used as a energy source for generating kinetic energy of large turbines rotated by using bladed fans. This kinetic energy is converted in to electric energy. It covers 2.6 % of total global power. But expect to cover 18 % in future suggested by reports in International energy agency (Al-Nouman *et al.* 2011). The initial cost for installation of wind mills are very high and some reports suggests these wind mills can affect ecosystem and biodiversity of nature mainly for aquatic life in sea.

Solar Energy

Sun being the largest source of energy for the planet earth and is considered as the largest source of clean renewable source of energy. Out of all these energy sources, solar energy is a clean source of energy. The total solar energy reaching earth surface is sufficient to meet whole world energy demand. Solar energy can be considered in to two variants namely light and heat. Solar photovoltaic is harvesting energy from electromagnetic radiations (from ultraviolet (UV) to infrared (IR) regime) and converting in to electric energy. Solar thermal is harvesting energy from heat energy. Solar spectrum extends from UV to IR region with non-uniform intensity distribution extending in a wavelength range of 3 to 0.2 μm. A typical solar spectrum is shown in Figure.

1.1. From the spectrum it can be seen that area under the curve gives the total intensity distribution of ˜ 1.35KWm$_2$ (Rephaeli *et al.* 2009).

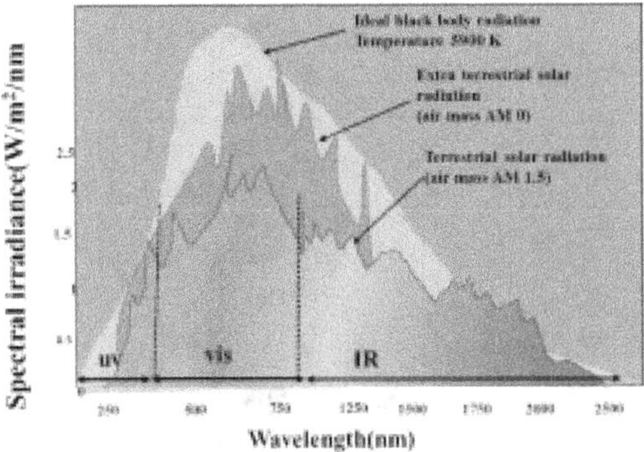

Figure 1.1 Solar Radiation Spectrum

PHOTOVOLTAICS

Photovoltaic (PV) is a promising renewable energy technology and best alternative for conventional energy sources. History of PV is backed by discovery of photovoltaic effect in 1839 by French physicist Edmond Becquerel. While experimenting with electrolytic cell made of two metal electrodes it was discovered that on exposure to light certain material can regenerate small amount of current. From that period onwards various devices and technologies have been developed with the objective of developing renewable energy source by converting sunlight to electricity. Over the last two decades, the PV market is the fastest growing industry for its size, maintaining a high growth rate of more than 40 %, making it the fastest growing energy conversion technology. Figure 1.2 gives the PV cell development history and its broad classification based on the underlying technology (De Brito *et al* 2011).

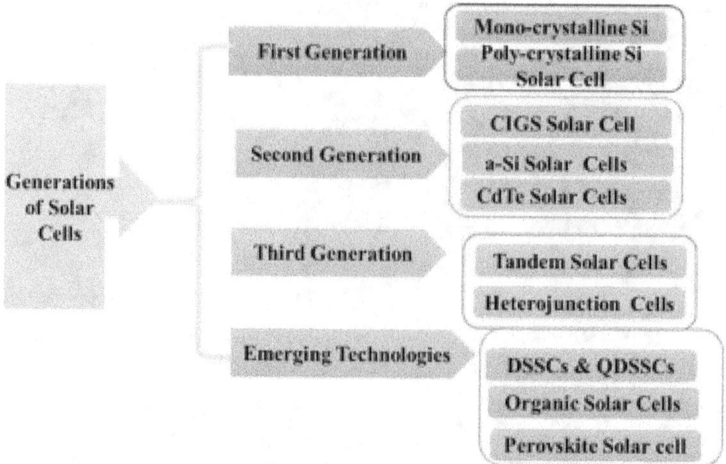

Figure 1.2 Different generations of Solar cells

Generations of Solar Cells

Single and multicrystalline p–n junction silicon cells are the most common and dominant PV of the solar cell market with efficiency of nearly 25

% and market share of about 85 %. These cells offer best reliability with least efficiency degradation (life time more than 25 years) and are called first- generation devices. High purity requirements of the silicon (Si) crystals, high fabrication temperatures and large amount of material needed for a wafer based cell are major limiting factors. Sophisticated fabrication techniques and complicated material processing of cell makes it very expensive.

Efforts to reduce manufacturing cost and recombination losses leads to emergence of second generation PV using thin film technology. Thin layers of inorganic semiconductors like Copper Indium Galium Selenide (CIGS), Cadmium Telluride (CdTe), Copper Indium diSelenide (CuInSe$_2$) thin films or amorphous and nanocrystalline silicon deposited between a transparent conducting substrate and a back electrode form thin film technology based PVs.

They have current market share of about 15%, with relatively lower efficiency (~20%) compared to first generation solar cells. However, these PV panels have reached the stage of commercialization and have entered the PV market. Moreover, the thermodynamic limit of the light to electric power conversion efficiency of a single-junction PV cell (1st or 2nd generation) under AM 1.5G spectrum is about 33 %. This limit is known as Shockley–Queisser limit originates from the fact that photons with energies below the bandgap energy are not absorbed, while photons with energies above the bandgap energy release the additional energy ($E_{photon\_}E_{gap}$) mostly as thermal energy.

Technology that came up to develop PV with conversion efficiency beyond Shockey-Queisser limit through advanced PV concepts such as multijunction solar cells, optical up- and down converters, tandem solar cells leads to the development of third generation solar cells. The complicated and expensive manufacturing process of first generation solar cells and low conversion efficiency of second generation solar cells are overcome in the third generation solar cells. Though third generation cells showed efficiency above 45%, the processing of these cells remained to be highly expensive. Due to these derelictions of aforementioned solar cells, a big challenge is to develop new, advanced materials with good optical and electrical properties that can be highly efficient, mass-producible and inexpensive.

Harvesting energy from sun by using nanotechnology based novel approaches –came up a new era of PV technology named as a recent emerging PV technology which includes dye sensitized solar cell (DSSC), organic solar cell, quantum dot sensitized solar cell (QDSSC), and perovskite solar cells (PSC) Della *et al.* (2015), Lee *et al.* (2009), Li *et al.* (2014). Though these cells are recent emerging PV technology, they have shown an excellent progress in efficiency within very short span of time. Sensitized solar cells possessed inexpensive fabrication process and potential to achieve high efficiency up to

44.7 % by overcoming Shockley–Queisser thermodynamic limit (Choubey *et al.* 2012). In 1991 O'Regan and Gratzel introduced DSSC with optically active sensitizing material such as Ru based dye sensitized on mesoporous photoanode, liquid electrolyte and CE (Roy *et al.* 2010, Gratzel *et al.* 2005, Prabavathy *et al.* 2017). Owing to its light weight, flexibility, low toxicity and good performance at various optical condition DSSC made considerable attention among various PV cells. But dyes were limited by its single wavelength absorption and high cost. Inorder to overcome these limitations, dyes were replaced by new type of light absorbing material i. e. quantum dots (QDs) for designing QDSSC (McDaniel *et al.* 2013).

Quantum Dots

QDs are 3D confined nanoparticles with dimensions less than or equivalent to its excitonic Bohr radius. The Bohr radius is calculated from the following expression as shown in (eqn 1.1).

r = (EmO)r

—

(1.1)

e m* 0

where Є is the material's relative dielectric constant, m_0 is the free electron mass, r_0 is the Bohr radius of Hydrogen atom (˜ 0.529A) and m_* is the reduced effective mass given by Equation.1.2.

$m_* = 1 + 1$

me mh

(1.2)

where m_e and m_h are the mass of electron and hole respectively.

The second important parameter explaining QDs behaviour is De Broglie wavelength. When the dimension of semiconductor nanocrystal approaches the De Broglie wavelength of the charge carriers (electron and holes) in the bulk semiconductor, the quantum confinement effect occurs.

De Broglie wavelength is given by Equation. 1.3

$$\underline{\hspace{2cm}} \Lambda = h\,m_{eff}kT$$

(1.3)

where h is Plank's constant, m_{eff} is the electron's effective mass, k is the Boltzmann constant and T is the absolute temperature.

Quantum confinement effect leads to energy band gap differences depending on the size of the QD, given by Equation. 1.4.

$$E =_1 \\ \underline{}g\,r2$$

(1.4)

where r denotes the QD's radius. When QD size decreases, band gap widens and hence more energy is required to excite small QDs than bigger ones. Thus, larger QDs have a red-shifted optical spectrum and smaller QDs have blue shifted spectrum.

Due to confinement effect, QDs exhibit unique opto-electronic characteristics like size dependent tunable band gap and multiple carrier excitation (MEG). The MEG process can convert a high-energy photon to multiple electron–hole pairs instead of a single pair in a bulk material, resulting inverse Auger effect for obtaining higher efficiency in QDSSCs. By utilizing these characteristics, overall performance of QDSSCs can be enhanced. Thus, QDSSCs are attracting enormous attention, and the number of reports increases annually (Jacak *et al.* 2013).

Significant Characteristics of Quantum Dots

◇ Tunable optical absorption

◇ Exciton generation potential

◇ High optical extinction coefficient

◇ Large intrinsic dipole moment

◇ Hot electron injection

◇ Low production cost

◇ Solution processability

Tunable Optical Absorption

QDs have energy separation between sub bands (various quantized levels for electrons and holes) much wider than its excitonic binding energy. This confinement effect allows tuning of optical and electrical properties of material by varying size and shape of QDs.

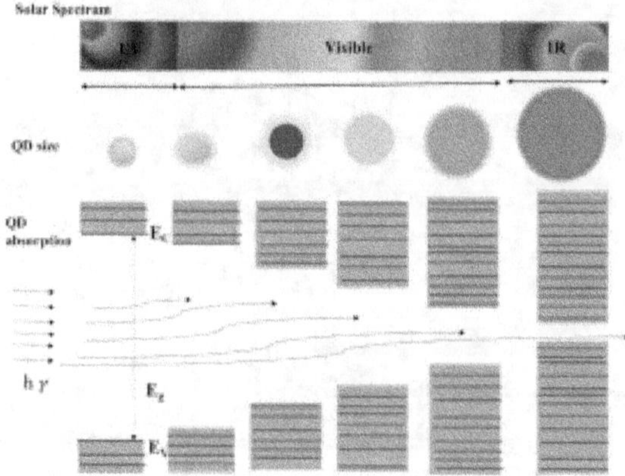

Figure 1.3 Size dependent tunable bandgap of Quantum Dots

As shown in Figure 1.3 the size dependent tunable band gap of QDs makes it possible to tune optical activity to a wider spectral window. Thus, by tailoring size of QDs and combining different size QDs improved light absorption can be achieved which leads to improve cell efficiency (Zhao *et al.*

2017).

Multiple Exciton Generation (MEG)

MEG in QDs can be defined as generation of multiple excitons upon absorption of a single photon. Upon absorption of solar radiation, the incident photon energy in excess of the band gap is lost as thermal energy as heat through electron–phonon scattering. But in QDs excess threshold photon energy is used for impact ionization.

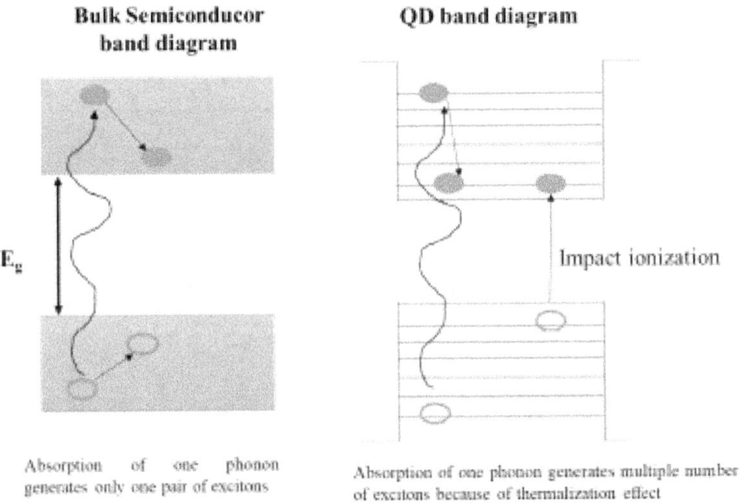

Figure 1.4 Band Diagram on multiple exciton generation in Quantum Dots

The excess kinetic energy is equal to the difference between the photon energy and the bandgap, which creates an effective temperature condition in carriers as hot carriers. Hot carriers cause impact ionization in QDs generating additional excitons. The energy higher than band gap stimulates

impact ionization in QDs generating additional excitons. The energy higher than band gap stimulates semiconductor materials to generate another pair of excitons. thus, energy is conserved (Beard *et al.* 2008). Schematic representation of multiple exciton generation is shown in Figure 1.4.

QUANTUM DOT SENSITIZED SOLAR CELLS

QDSSC is one of the emerging PV technologies that uses semiconductor QDs as a sensitizing material for absorbing solar light. QDSSC have attracted significant notability for easy fabrication and a low cost. Cd- and Pb based QDs are widely studied sensitizer materials for QDSSC. The significant properties of QDs like tunable bandgap, MEG and hot carrier generations provides promising way for QDSSC to improve photovoltaic efficiency. QDSSCs attracted significant interest in comparison to DSSC due to its simple cell fabrication, cost effectiveness and enhanced light harvesting capability achieved using QDs. However, QDSSCs structure and working principle reveals that they are similar to DSSC, although the variation comes with the sensitizer part. Though it possesses aforementioned significant properties the efficiency of QDSSC is still lower than other emerging solar cell technologies like DSSCs, perovskite solar cells and polymer solar cells. The efficiency of QDSSC is greatly affected by type of sensitizer (QDs) their size, morphology, interface stability between QD & metal oxide nanostructures (Jun *et al.* 2013b). Further systematic investigations are needed to further improve the efficiency of QDSSC. Basic schematic diagram of QDSSC is shown in Figure 1.5.

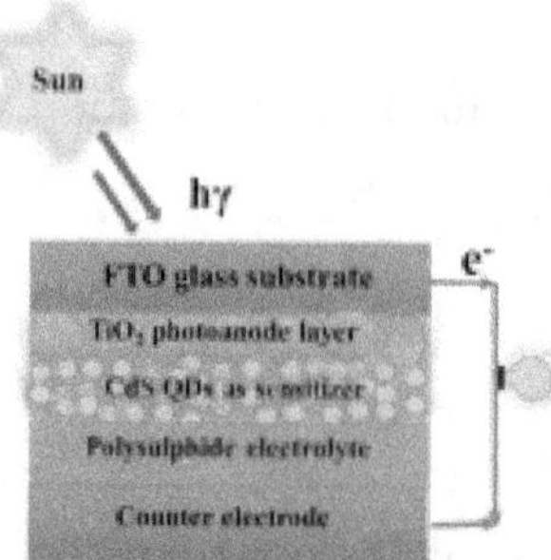

Figure 1.5 Schematic Diagram of Quantum Dot Sensitized Solar Cells

COMPONENTS OF QDSSC

The main components of QDSSC includes semiconductor oxide based photoanode, semiconductor QDs as sensitizer, electrolyte containing a redox couple and CE. The role of each component is highly significant for the overall performance of QDSSC. The materials used for each component and their impact on the performance of QDSSCs are discussed in the following sections.

Photoanode

Wide band gap oxide semiconductors like TiO_2, ZrO_2, In_2O_3, ZnO, SnO_2, Nb_2O_5 coated on to transparent conducting oxide (TCO) substrates are commonly used as photoanodes. They are also known as working electrode which absorbs sensitizer material and collects the excited electrons from sensitizer and transport to the conducting substrates. The band gap, morphology and composition as well as the thickness of photoanode layers influence its charge collection efficiency which results the variation in the performance of QDSSCs.

Fluorine-doped tin oxide (FTO) or indium tin oxide (ITO) are commonly used TCO substrates for photoanodes of QDSSC. ITO substrate exhibits transmittance over 80 % and resistivity of 10_{-4} Ω cm at room temperature and hence commonly used in optoelectronic applications. But the low resistivity of ITO substrate could deteriorate during the high temperature treatment in the QDSSC fabrication. Fabrication of photoanodes of QDSSCs involves coating of metal oxides on the conductive face of ITO substrates followed by high temperatures calcination at 450 ∘C, for improved adherence and conductivity. But when ITO films are treated with temperatures above 300 ∘C, the sheet resistance increases, resulting poor performance. Whereas, FTO substrates showed nearly 70 % transmittance in the visible range at a thickness of 750 nm, which is nearly 10 % lower than ITO substrate. The sheet resistance of the FTO substrate is ~ 12 Ω which is 20 % less than that of an ITO substrate after calcination at 450 ∘C. In addition, it can be seen that the cost of FTO is one-third less than that of ITO substrates. Therefore, FTO coated substrate is preferred over ITO substrates for the fabrication of photoanodes.

Many studies have been reported on different wide band gap oxide semiconducting materials like TiO_2, SnO_2, Zn_2GeO_4, Nb_2O_5, ZnO etc. TiO_2 having bandgap of 3.2 eV and it is an efficient photoanode material for QDSSC. Table 1.1 shows performance of QDSSC using different photoanode materials. Many studies reported synthesis of TiO_2 nanotubes by hydrothermal method and studied effect of various parameters like temperature, pH in QDSSC performance. Kim *et al.* (2015c) studied the effect of branched TiO_2 nanorods in Mn doped CdS and CdSe QDs based QDSSCs. The study focussed on effect of Mn doping in improving photovoltaic performance of QDSSCs along with the role of morphology of TiO_2 nanostructure. The study highlighted that rod like structure of TiO_2 suppressed charge-transfer resistance and thereby enhancing electron transport and carrier lifetime. Another study by Ruhle *et al.* (2010), reported on suppression of recombination at FTO/ electrolyte interfaces achieved using TiO₂ compact layer as photoanode. TiO₂ is widely used as photoanode material for solar cell because of its easy availability, low cost, biocompatibility. simple synthesis procedure and stability. ZnO is next choice as a photoanode material for QDSSC which is having band gap of 3.37 eV.

Several works were also reported for improving QDSSC performance using TiO₂/ZnO photoanodes (Soel *et al.* 2010). From Table 1.1 the variation in QDSSC performance by using different TiO₂ nanostructures can be observed. It is clear from the Table 1.1 that structural and morphological properties of photoanode material have a significant influence on enhancing the efficiency of QDSSC. Hence it is highly imperative to synthesis the photoanode material with controlled morphology in order to provide larger surface area for absorption of QDs and to improve charge diffusion pathway for transport of injected carriers to external circuitry and thus to enhance the QDSSC performance.

Table 1.1 QDSSC using different Photoanode material.

QDs	Photoanodes	J_{sc} (mA. cm^{-2})	V_{oc} (V)	FF	Efficiency (%)	Reference
CdS	H-TiO$_2$ NRA	4.95	0.500	0.42	1.04	(Liu *et al.* 2015)
CdS	TiO$_2$ NRs	7.13	0.66	0.49	2.31	(Gao *et al.* 2016)
CdS	TiO$_2$ paste	7.23	0.564	0.45	1.86	(Jeong *et al.* 2014)
CdS	TiO$_2$ NRs	3.68	0.59	0.49	3.14	(Wan *et al.* 2016)
CdS/ CdSe/ ZnS	TiO$_2$ MSs	13.89	0.540	0.54	4.05	(Wu *et al.* 2014)
CdSe/ ZnS/ SiO$_2$	TiO$_2$ NSs	16.95	0.591	0.50	5.01	(Zhou *et al.* 2017)
CdSe/ ZnS	TiO$_2$ NPs	15.54	0.563	0.61	5.53	(Du *et al.* 2014)
CdS/ CdSe/ ZnS	TiO$_2$ NTs	10.81	0.689	0.62	4.61	(Huang *et al.* 2013)
CdSe/ ZnS	TiO$_2$ NRs	17.5	0.563	0.60	5.96	(Zhang *et al.* 2016)
CdS/ CdSe/ ZnS	TiO$_2$ NFs-NTs	8.8	0.610	0.50	2.8	(Han *et al.* 2013)
CdS/ CdSe	ZnO NWs	17.3	0.627	0.88	4.15	(Seol *et al.* 2010)
CdS/ CdSe/ ZnS	TiO$_2$ MPs	14.05	0.571	0.51	4.15	(Du *et al.* 2016)

Sensitizers

Sensitizers should possess high absorption coefficient for getting maximum light harvesting efficiency. QDs are light absorbing material for QDSSCs. QDs are anchored on to photoanode by different methods like SILAR, CBD, DA etc (Ruhle *et al.* 2010). An ideal QD material upon exposure to sunlight must absorb photon and produce photoexcited electrons. Excited electrons should move on to photoanode and to conductive substrate at a very fast rate. For facile electron transfer the energy levels of QD must match with that of wide band gap material on photoanode (Yue *et al.*2018).

Different QD materials are used as sensitizer for QDSSC includes Sb_2S_3, Ag_2S, CdS, Bi_2S_3, CdSe, Ag_2Se, CdTe, CdS/CdSe, $CuInS_2$, Cu_2S, PbS,

$Pb_xCd_{1-x}S$, PbSe, graphene, carbon etc (Duan *et al.*2015, Yuan *et al.* 2016). Table 1.2 summarizes the cell parameters of QDSSC using different sensitizers and co-sensitizers. Cd based QDs are most widely used and investigated as they are more stable and easier to synthesis in comparison to other QDs. For improving light absorption range beyond visible range of solar spectrum co-sensitization of QDs is an efficient approach. CdS/CdSe, CdS/PbS, CdS/CdTe, CdS/InSb, In_2S_3/PbS are the reported co-sensitizers for QDSSCs (Lee *et al.* 2009; Mu *et al.* 2013; Bai *et al.* 2016). Broadened light absorption range extending up to near IR regime with reduced recombination rate can be achieved by co-sensitization of QDs which improves performance of QDSSCs. Yang *et al.* (2016) highlighted the improved efficiency of 7 % achieved with CdTe/CdSeS co-sensitization. Additionally, the role of passivation layer in improving the QDSSC cell performance can also be inferred from Table 1.2.

Table 1.2 QDSSC using different sensitizer material and poly sulphide electrolyte

Photoanode	QDs	CE	Efficiency (%)	Reference
TiO_2 MSs	CdS/CdSe	Pt	4.81	(Yu *et al.* 2011)
TiO_2 NPs	Ag_2S/ZnS	Pt	1.76	(Wang *et al.* 2015)
TiO_2 NPs	CdS/CdSe/ZnS	gold	4.22	(Lee *et al.* 2009)
TiO_2 NPs	$CuInS_2$/ZnS	Cu_2S	7.04	(Kim *et al.* 2015a)
TiO_2 NPs	CuInSe/ZnS	Cu_2S	8.1	(Bai *et al.* 2016)
TiO_2 NPs	ZCIS	Cu_2S/brass	8.47	(Yue *et al.* 2018)
TiO_2 NPs	CdTe/CdSeS	Cu_2S/brass	7	(Yang *et al.* 2016)
TiO_2 NPs	ZnCuInSe/ZnS	Carbon	11.61	(Du *et al.* 2016)
TiO_2 NPs	CdS/Bi_2S_3/ZnS	Cu_2S	2.52	(Mu *et al.* 2013)
TiO_2 NPs	$CuInSe_xS_{2x}$/ZnS	Cu_2S	5.51	(Mc Daniel *et al.* 2013)

Electrolyte

An electrolyte conducts electricity by movement of ions. An effective redox electrolyte should be highly soluble with high ionic mobility, electron transfer kinetics, thermally and chemically stable in both dark and under illumination. Polysulphide electrolyte is commonly used for QDSSC application. Main limitation on using aqueous polysulphide electrolyte is lower value of V_{oc} and poor FF because of high regeneration rate of QDs (Jun *et al.* 2013a; Du *et al.* 2015). In addition, some undesirable intrinsic properties of electrolyte that needs to be minimized includes photodegradation, leakage, volatilization of liquid solvents. These characteristics of liquid electrolyte greatly affect the long-term stability and commercialization of QDSSC. To overcome these limitations, several works have been focused in replacing liquid electrolyte with quasi-solid or solid electrolytes (Yu *et al.* 2010).

Also, electrolyte performance is enhanced by varying the concentration of redox mediator and by adding some additives like PEG, PVP, guanidine thiocyanate, SiO_2 etc (Du *et al.* 2015; Yu *et al.* 2017). So far on replacing liquid electrolyte with gel electrolyte showed only poor performance of QDSSCs. The recent reports showing significant improvement in QDSSC performance by electrolyte modification are listed in Table 1.3. In a study reported by Duan *et al.* (2015) on synthesis of succinonitrile and sodium sulphide based plastic crystal as a solid-state electrolyte by blending approach. Feng *et al.* (2016a) attempted to use sodium polyacrylate to gelate conventional polysulphide electrolyte. Dang *et al.* (2017) proposed a highly stable electrolyte using benzimidazolium salt maintaining 67 % of efficiency for long term (504

h) operation.

Counter Electrodes

The CE is an important part of the QDSSCs as it transfers electrons from the external circuit into the electrolyte and catalyses the reduction reaction of the oxidized electrolytes at the electrolyte/CE interface. The performance of CE mainly depends on conductivity of material, its chemical and mechanical stabilities. An ideal CE material needs to display excellent electrocatalytic activity towards reduction reaction of oxidized electrolyte and good stability at the CE/electrolyte interface (Murakami *et al.* 2008). Platinum (Pt) CE is widely used for DSSCs because of its stability and superior catalytic activity toward triiodide reduction. But Pt CEs are poorly active in a polysulfide electrolyte due to the chemisorption of sulfur species on the Pt surface. This causes large charge transfer resistance (R_{ct}) at the CE/electrolyte interface and a low current density (J_{sc}) and Fill Factor (FF).

Recently, as an alternate to Pt CE, significant efforts have been made to develop various CE such as conducting polymers (e. g., polyaniline, polypyrrole, and poly(3,4-ethylenedioxythiophene), noble metals (e. g., Au, Ag), metal chalcogenides)(e. g., copper sulfide (CuS), cobalt sulfide (CoS), nickel sulfide (NiS), lead sulfide

(PbS) and CuSe, FeS_2, Cu_2SnS_3, $Cu_2ZnSnSe_4$, carbonaceous materials (e. g. mesoporous carbon, carbon nanotubes, activated carbon, graphene, carbon black, and carbon foams) composite materials CuS/CoS, $NiCo_2S_4$ and CoS-NiS (Faber *et al.* 2013; Duan *et al.*2015, Xu *et al.* 2014). Among them, metal chalcogenides are most promising as they exhibit excellent electrocatalytic activity toward the polysulfide electrolyte with good physical and chemical stability, ease of fabrication, low cost (Feng *et al.* 2016b; Yu *et al.* 2010).

CuS is a p-type semiconductor with a band gap of 1.1–1.4 eV. CuS is widely used as CE in QDSSC and is generally fabricated by immersing brass foil in HCl to remove Zn component and to form Cu_2S. But as synthesized Cu_2S undergoes corrosion because of polysulphide electrolyte. Hence as an alternative method other CBD, SILAR etc have been tried on FTO substrate. NiS is another metal sulfides that has attracted great attention because of its remarkable electronic conduction, low toxicity, cost-effectiveness, simple fabrication process, and variety of valence states. Different phases of nickel sulfides such as, α-NiS, Ni_3S_2, β-NiS, Ni_3S_4, and NiS_2 are inexpensive and abundant materials with widespread applications in QDSSCs. An effective conversion route for the controllable synthesis of uniform NiS nanostructures includes in-situ chemical ion-exchange method.

Table 1.3 QDSSC using different electrolytes and counter electrodes

QDs	Electrolyte	CE	V_{oc} (V)	J_{sc} (mA/cm^2)	FF	Efficiency (%)	Reference
CdSeTe	(PAAS) gel	Cu_2S	0.66	20.63	0.62	8.54	(Feng et al. 2016)
CdS/CdSe	Polysulphide	CoS_2	0.51	14.44	0.56	4.16	(Fabe et al. 2013)
CuInS2/CdS	Polysulphide	PbS	0.58	18.3	0.45	4.70	(Lin et al.20..)
CdSeTe	(CMC-Na)	Cu_2S	0.66	21.89	0.63	9.21	(Feng et al. 2016)
ZnSe/CdSe	Polysulphide	FeS_2	0.743	13.58	0.38	3.9	(Xu 2014)

Table 1.3 (Continued)

QDs	Electrolyte	CE	V_{oc} (V)	J_{sc} (mA/cm^2)	FF	Efficiency (%)	Reference
CdSe	Polysulphide	CZTSe	0.54	15.49	0.52	4.35	(Zeng et al. 2013)
CdSe/CdS	PEGDMEh–Na2S–S	Carbon	0.72	17.84	0.42	5.45	(Kim et al. 2014)
CdS/CdSe/ZnS	Polysulphide	PbS	0.56	15.11	0.40	3.49	(Wang et al. 2018b)
ZnO/CdS	Polysulphide	NiS	0.60	7.14	0.56	2.5	(Han et al. 2017)
CdS/CdSe	PAM-MBAf	Cu2S	0.61	12.8	0.55	4.30	(Yu et al. 2010)
PbS–CuS	P3HT	Au	0.60	20.7	0.65	8.07	(Park et al. 2016)
CdS/CdSe	Polysulphide	MoS2	0.58	15.65	0.44	4.13	(Wu et al. 2017)
CdS/CdSe/ZnS	Polysulphide	CuInSe2	0.55	14.74	0.36	2.92	(Guo et al. 2018)
PbS/CdS/ZnS	Polysulphide	Cu2S	0.44	19.6	0.48	4.24	(Wang et al. 2018a)

PbS is a direct gap semiconductor with a band gap of 0.41 eV in the bulk state belongs to IV–VI group. PbS emerged as a low-cost p-type absorber layer material for PV applications because of its stability and low toxicity compared to other lead chalcogenides. Larger exciton Bohr radius of 18 nm, leads to strong quantum confinement resulting the optical absorption band from

0.4 to 1.5 eV, which results in large relativistic splitting in the band structure of PbS, making it favourable for band gap engineering (Duan *et al.* 2015).

WORKING MECHANISM OF QDSSC

Schematic representation of QDSSC is shown in Figure 1.6. The photoanode is fabricated by coating a mesoporous wide bandgap semiconductor layer (TiO_2) with an optimal thickness of 10 μm on to the FTO substrate. The QD photosensitizers are loaded onto the surface of photoanode by SILAR method. The QD loaded photoanode and CE are sandwiched to fabricate a QDSSC. The electrolyte is filled in between the photoanode and CE.

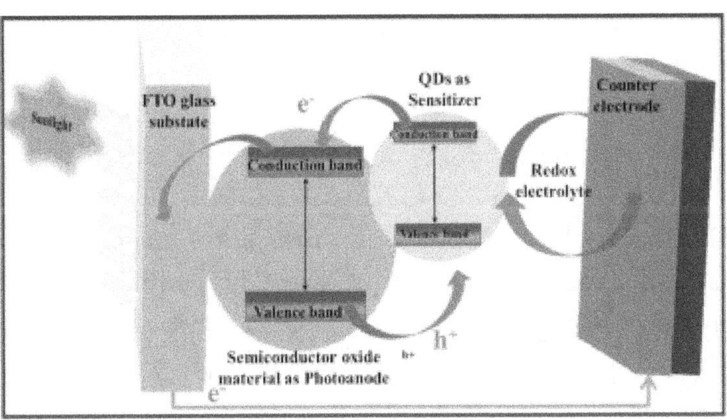

Figure 1.6 Schematic diagram for QDSSC configuration

QDSSCs operation involves the following process:

◈ On persistently irradiating QDSSC by incident light, the QDs can absorb photons and excite electrons from valence band of QDs to their conduction bands thereby producing electron hole pairs at the interface between the nanocrystalline TiO_2 and QDs.

◈ The excited electrons from the QDs are injected into the conduction band of TiO_2 and then transferred through TiO_2 to FTO while the holes move to the polysulfide electrolyte.

◈ QDs are subsequently regenerated by redox couple reaction of the polysulfide electrolyte through electron donation.

52

◇ The electrons reach CE via an external circuitry. The sensitizer layer accepts electrons from redox mediator and oxidized redox mediator diffuses through CE.

A major force that counteracts the QDSSC operation processes is the charge recombination of electrons in the conduction band of QDs and TiO_2 and also QDs with that of electrolyte interface (Jun *et al.* 2013b).

Several studies were reported on controlling recombination losses for improving the performance of QDSSC (Rao *et al.* 2015, Duan *et al.* 2015 & Lee *et al.* 2017). Several parameters influence the performance of QDSSC such as the short-circuit current density (J_{sc}), open-circuit voltage (V_{oc}), fill factor (FF), are shown in plot in Figure 1.7. Under one sun illumination, the efficiency (η) of a solar cell can be obtained using the following Equation.1.5.

$$\eta\,(\%) = \frac{*}{P} = \frac{e* \; e*}{P} *100 \quad (1.5)$$

where P is the incident solar power on the device and J_{sc} depends on light harvesting, charge collection and electron and hole injection efficiencies. The decay rate of V_{oc} is directly related to the carrier lifetime. Also, V_{oc} is the maximum voltage that a solar cell can generate which is determined by the difference between the Fermi level of the wide band-gap semiconductor with respect to the redox potential of the redox system (Kusuma *et al.* 2018). FF is the quality determining parameter of cell and can be defined as ratio of product to maximum voltage and current to product of V_{oc} and I_{sc} is given in Equation 1.6.

$$FF = \frac{*}{e* \; e}$$

$$(1.6)$$

where V_m is maximum voltage and I_m is maximum current.

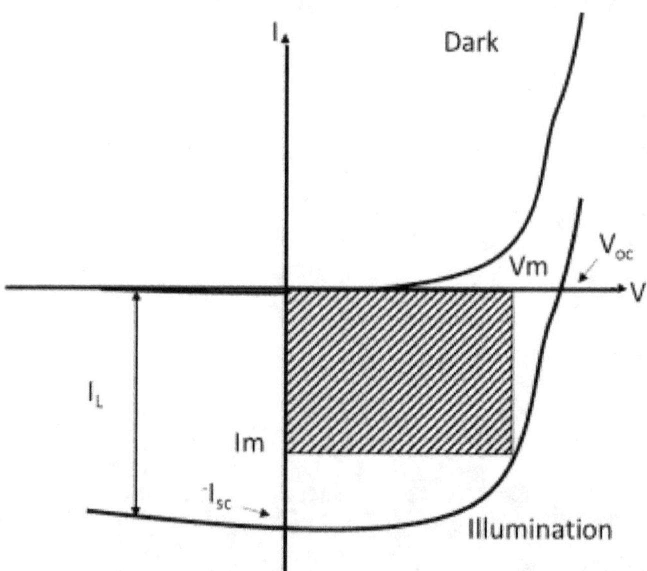

Figure 1.7 J-V plot of solar cell with maximum power as indicated by shaded area in the plot. V$_m$ and I$_m$ are corresponding maximum voltage and current

QDSSC FABRICTION TECHNIQUES

Techniques used for fabricating QDSSCs are important area for optimization to enhance solar cell performance. Commonly used methods for preparing photoanodes, deposition of QDs and CEs are summarized below.

Preparation of Semiconductor Photoanode

A thin layer of wide band gap semiconductor oxide material is deposited on conductive FTO substrate through different types of deposition methods like doctor blading, screenprinting, spin coating, hydrothermal, sputtering etc. For improved performance of QDSSC device, a porous large surface area thin film having excellent loading capacity of QDs is required. Among aforementioned thin film deposition techniques, doctor blading is a simple and well known method. Hence this method is widely used for photoanode fabrication.

(i) Doctor blading and screenprinting- techniques

In doctor blading, uniformly dispersed nanoparticles are made using a viscous paste by mixing with appropriate ratio of an organic binder (e. g. ethyl cellulose), surfactant (e. g. terpineol) and solvent (e. g. ethanol). By using a scotch tape or mask an area is selected in FTO substrate on which the nanoparticle paste is dropped. A thin uniform film of semiconductor coating is made by spreading the paste using a sharp blade. As prepared thin film coating is dried at room temperature and then annealed at high temperature for a fixed period. Temperature, composition of paste, speed and angle through which blade is moved on substrate to spread paste are significant factors that determine the film quality. However, better control on film thickness can be done using multilayers of scotch tape for masking the deposition area (Gratzel *et al.* 1994). Doctor blading is a simple technique but reproducibility of same film quality is always a challenge in this technique.

While, screen-printing technology have added advantage to reproduce thin films with same quality at the same time useful for large scale production. Hence it is widely used method for commercial production of conventional silicon solar cells. In screen printing technology, the conductive surface of substrate is placed with a mask having mesh type structure. Prepared paste (similar as prepared for doctor blade technique) is dropped on mask which fills all the open area of mesh. Later, paste is applied through open area of mesh using a machine operated squeegee. The residual paste is expelled out from the surface of electrode. After removing the mask and as prepared semiconductor material coated layer is dried and annealed. Sufficiently large area of uniform thin film deposition can be done by using screen-printing method (Du *et al.* 2014, Liu *et al.* 1993).

(ii) Spin Coating method

In spin coating technique semiconductor thin film is prepared by dropping the paste in the center of conductive surface of FTO substrate. While at same time the conductive glass substrate is spun at a very high rpm through which it assists the spreading of paste throughout the substrate. The residual amount of paste is thrown away from the substrate. Prepared film is then dried at room temperature followed by annealing at high temperature for a defined time. The coating process is repeated number of times for required appropriate thickness of thin film. Thin film characteristics depends on paste composition, drying temperature, and speed of spin coating unit (Dinh *et al.* 2011).

(iii) Hydrothermal Method

Hydrothermal method is a technique for thin film deposition in which reaction takes place at controlled high temperature and pressure. The hydrothermal process includes a stainless steel covered vessel called autoclave within which a teflon lined inner vessel. The substrate is immersed into the aqueous precursor solution filled in the teflon vessel. Film growth depends on temperature, pressure and time of reaction set for sealed autoclave. Hydrothermal is a facile one step synthesis process which provides uniform size distribution. If solvent other than water is used, then the method is referred as solvothermal technique (Vijayalakshmi *et al.* 2012).

(iv) Other Methods

Other techniques involved for thin film deposition includes sputtering, spray pyrolysis, electrospinning, atomic layer deposition and so on. By combining two techniques fabrication of hierarchical structures can be done. For example, to prepare 1D structures like nano rod, nanotubes, seed layer is initially deposited by spin coating or sputtering technique while further growth of 1D nanostructure is done by hydrothermal technique. To fabricate photoanode with mesoporous spherical morphology, spin coating /spray pyrolysis is suggested for seed layer deposition while hydrothermal method is more effective for synthesizing vertical 1D structures (Hasegawa *et al.* 1993).

Preparation of Quantum Dot sensitizer layer

In-situ or ex-situ method can be used for depositing QD layer on semiconductor coated photoanode.

(i) In-situ methods

Chemical bath deposition (CBD) and successive ionic layer adsorption and reaction (SILAR) are most commonly used for QD deposition through in-situ approach. These methods are cheap and low cost methods that can even be used for large scale production (Samadi Maybodi *et al.*2016).

(a) Chemical Bath Deposition

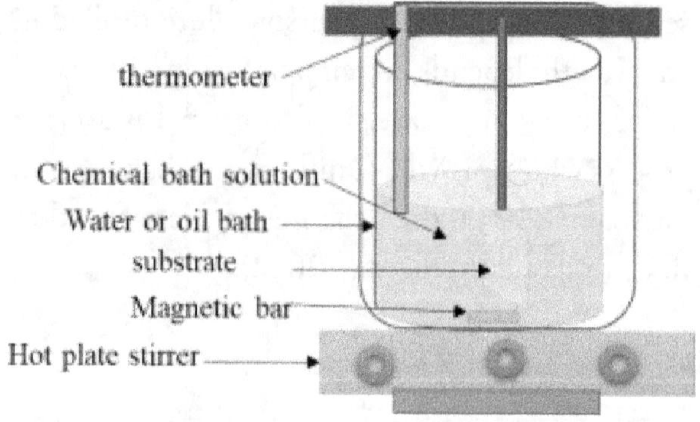

Figure 1.8 Experimental set up for Chemical Bath Deposition

In CBD technique, chemical deposition is carried out in one bath as shown in Figure 1.8. Initially anionic and cationic precursor solution are prepared separately and made to react in one container named as bath solution. Photoanode material is dipped into bath solution for a specified period and slow reaction takes place for QD nucleation and growth. By varying and controlling the dipping time, precursor solution concentration, and temperature the deposition can be optimized (Chang *et al.* 2007). Liu *et al* (2010). used CBD method for CdS QDs deposition process and observed effect of temperature on thin film formation.

Liu *et al.* (2010) reported use of CBD method for CdSe QD deposition over TiO_2 thin film and observed improved photoelectrochemical characteristics of $CdSe/TiO_2$ assembly. CBD gives dense deposition of QDs on wide band gap semiconductor layer and enhances the recombination resistance. Also, nonuniformity in QD deposition is a problem while it can result in nanocrystalline film pattern which can block the narrow pore channel for pores having size in the order of 5–7 nm. Hence, CBD is suitable for nano-structured electrodes which requires large porosity of photoanode layer like nano-wires, nano-rods or other higher order hierarchical structures. Mainly due to above discussed shortcomings, further modification is done in CBD technique for fabrication of efficient and stable QDSSC. Another approach includes, microwave assisted CBD (MACBD) for rapid nucleation and growth of QD on TiO_2 deposited photoanode. CdS/CdSe co-sensitized TiO_2 photoanode is an example for fabrication through MACBD. This approach suppressed carrier recombination and hence, improved the efficiency (3.06 %) of QDSSC. Microwaves assists in quick and homogenous heating to the bath solution which results in faster nucleation and growth of QDs. MACBD also can enhances the wettability of photoanode surface which results in good contact between QDs and photoanode layer.

(b) SILAR method

SILAR is different from CBD method in which cationic and anionic precursor solutions are prepared separately and kept in two different containers. Semiconductor coated photoanode is first dipped into cationic precursor followed by rinsing and drying. Then, is dipped into anionic precursor, again followed by rinsing and drying. The whole process of sequential dipping of semiconductor coated electrode into cationic and anionic precursor solution each time followed by rinsing and drying is referred as one SILAR cycle (Figure 1.9). The desired QD growth depends on number of SILAR cycles, rate of dipping, dipping time, reaction temperature and composition of precursor solutions.

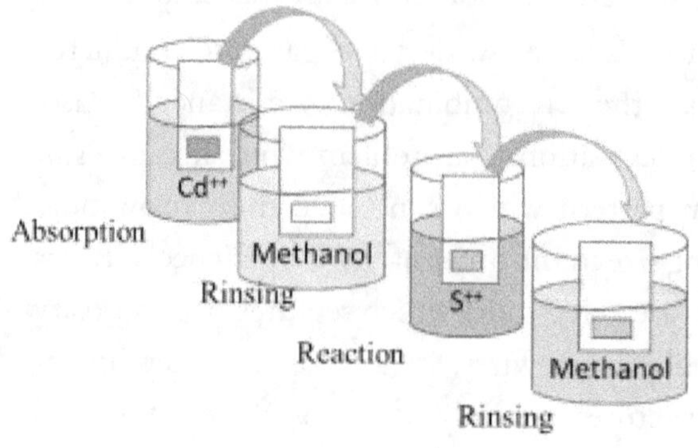

Figure 1.9 Schematic representation of SILAR process

A comparison on SILAR and CBD shows that SILAR has less processing time and maintain close stoichiometry formation. Recently, a new modification in SILAR named as Potential Induced Ionic Layer Adsorption and Reaction (PILAR) has been evolved (Lokhande *et al.*1991). In PILAR process, adsorption of cations was done under an applied bias of electrochemical cell followed by rinsing and drying. Afterwards, it was dipped into anionic precursor solution for completion of reaction and adsorption of QDs (Ha *et al*

.2015).

(c) Electro Chemical Deposition (ECD)

ECD technique, is a facile, low cost technique for producing semiconductor film of desired thickness. In the ECD method (Figure 1.10) the substrate and reference substrate are immersed into a bath solution consisting salt precursor. Thin layers of semiconductor are grown on substrate and is carried out in presence of electric field. On applying electric current, positively charged ions move towards negatively charged substrate. Thus, semiconductor thin film is deposited on substrate. Rao *et al.* (2014) have used ECD technique for deposition of CdS/CdSe QDs on TiO_2 nanowire array and achieved 4.20 % efficiency.

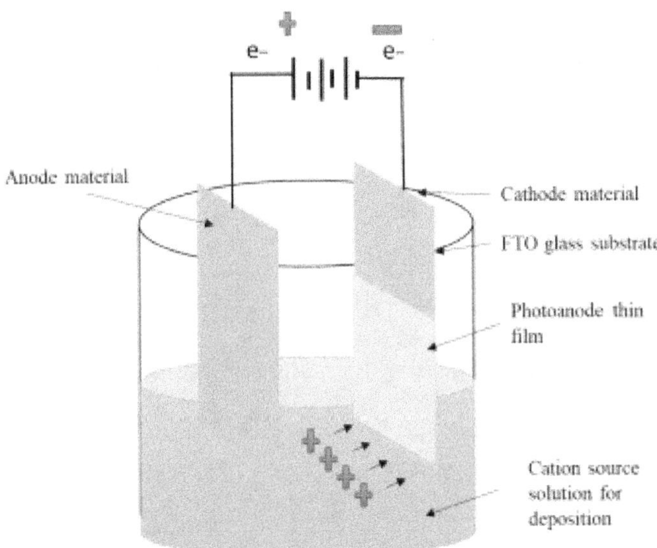

Figure 1.10 Schematic diagram of Electro Chemical

deposition process

(ii) Ex-situ methods

QDs are separately synthesized and is deposited on to semiconductor coated photoanode. The commonly used ex-situ techniques for QD deposition includes molecular linker attachment (MLA), direct adsorption (DA), and electrophoresis.

(a) Molecule linker attachment

In MLA a bi-functional linker molecule is mainly used for attachment of pre-synthesized QDs on to the photoanode surface. Bi-functional linker molecule have carboxylic acid group that is attached to semiconductor surface and on the other side of bi-functional linker molecule has a thiol group, that interacts to QDs. For QDSSC fabricated by MLA techniques, sensitized QDs will have its size, growth, structural, electrical and optical properties depend on concentration of capping agent in the precursor solution, reaction temperature and reaction time. In addition, electron transportation and the interaction between QD and photoanode will be based on length and electric properties of linker molecules (Pernik *et al.* 2011).

(b) Electrophoretic Deposition (EPD)

Electrophoretic bath is made by mixing polar/non polar solvent and colloidal solution of pre-synthesized QDs. Semiconductor coated photoanode as a positive terminal and FTO substrate used as a negative terminal. Both electrodes are kept at a specified distance and immersed into prepared electrophoretic bath. A DC voltage is applied for a fixed period to deposit QD on positive electrode (semiconductor coated photoanode). The quality of QD is precisely controlled by varying the concentration of EPD solution and deposition time in order to maintain size and shape of QDs (Salant *et al.* 2010).

(c) Direct Absorption

DA is an ex-situ approach in which the process is carried out by simple immersion of semiconductor photoanode material into colloidal solution for a period of time. The QDs get attached to photoanode surface through direct absorption without any external DC voltage or linker molecule. Presence of linker molecules can affect the effective charge transportation, leading to poor performance of solar cell. Therefore, this problem can be controlled by DA. But it suffers from a problem of nonuniform distribution of QDs on photoanode surface which has to be further optimized prior to its application. Proper deposition method can be selected depending upon the objective of specified investigation (Pernik *et al.* 2011).

(d) Other Methods

QDs can be deposited by some other methods like chemical vapour deposition (CVD) and spray pyrolysis. Spray pyrolysis technique combines both solution precipitation and gas phase processing. Initially formation of precursor aerosol droplets delivered through carrier gas maintained at a heating zone. Metal acetates, metal chlorides and metal nitrates are commonly used as precursor solutions. These precursor solutions are atomized into fine droplets and sprayed into the heated zone. Inside the thermal zone, the solvent gets evaporated and reactions starts to occur within each particle forming final product. One of main advantage of spray pyrolysis is its ability to form multicomponent nanoparticles as different metals can be mixed in a solution form and these solutions can be aerosolized into reaction chamber. Commonly dense particle in size range from 100 to 1000 nm can be prepared in large volume using this method (Fazli *et al.* 2016).

CVD methods can be grouped as hot-wall or cold-wall processes. Hot-wall or thermal CVD that uses a high temperature tube furnace to which the substrate is loaded inside the tube. The whole tube is heated by the furnace in order to heat the substrate and to initiate the growth conditions. Cold-wall processes or plasma enhanced CVD only heats the sample by controlling the temperature of the sample holder, and by leaving the whole system at a relatively low temperature. Commonly used CVD is thermal CVD. In thermal CVD solid material deposited on heated substrate by vapor chemical reaction occurring inside in a heated chamber. Normally reaction chamber will be filled with inert gas. By varying reaction temperature, reaction time, reaction gas mixture composition, gas pressure inside chamber different materials can be deposited. Material in power form or as solid crystal or coating can be prepared using CVD method (Jeong *et al.* 2008).

Preparation of Counter Electrode

In QDSSC, for preparation of CE different methods such as CBD, SILAR, drop casting, spin coating, sputtering, hydrothermal, electrochemical deposition, vacuum deposition, successive ionic solution coating and reaction (SISCR) are used. Conductive substrate materials like FTO or ITO substrates or metal sheets with chemical processing and treatment are used as base material for CE. SILAR and CBD are commonly used for fabrication of metal chalcogenides based CE, while doctor blade method is used for fabrication of thin film of metal or metal oxides which can serve as a CE. Technical procedure of SILAR, CBD and doctor blade are similar to the procedure discussed in previous subsection.

STEPS INVOLVED IN FABRICATION OF QDSSC

The fabrication of QDSSC is schematically shown in Figure 1.11. The fabrication process consists of following steps:

i. Cleaning the FTO substrate.

i. Identifying conductive side of FTO substrate using a multimeter.

  i. With the conductive side of FTO substrate facing up and applying two parallel strips of scotch tape on the edges of the substrate.

  ii. Applying nano TiO_2 paste on the top edge of the FTO substrate in between the exposed area of adjacent scotch tape by doctor blading.

  iii. Sintering the TiO_2 coated FTO substrate at 450 °C for 30 min.

  i. Soaking the electrode with QD sensitizers (SILAR, DA).

  i. Preparation of the CE with metal sulfide /Pt coating on FTO substrate.

  ii. Sandwitching the two electrodes together with a layer of parafilm as a spacer for electrolyte filling.

  iii. Finally, the space between the photoanode and CEs were filled with electrolyte by capillary action and sealed with binder clips.

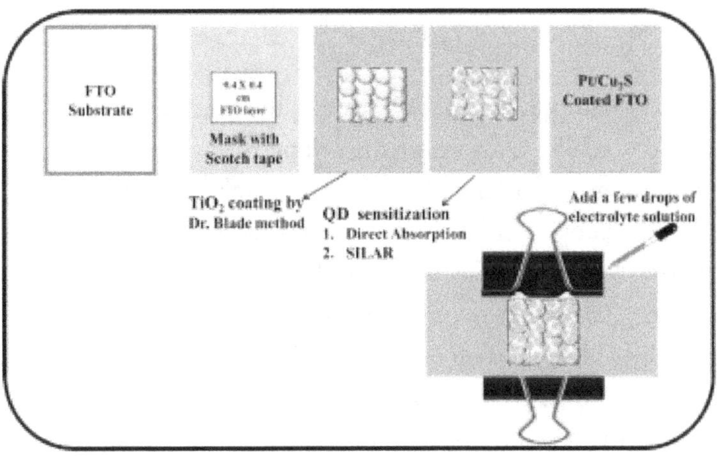

Figure 1.11 Procedure for fabrication of QDSSC

BRIEF LITERATURE SURVEY ON QDSSC

Yu *et al.* (2011) reported CdS/CdSe QDs prepared by electrodeposition in QDSSC and compared the variation of electron transport and charge recombination processes occurring at monosensitized and cosensitized QDSSC. Salant *et al.* (2010) reported on advantage of using electrophoretic deposition of QD using CdSe as QD material. Ning *et al.* (2014) reported the first photovoltaic-quality n-type colloidal quantum dot ink with an efficiency of 6% using colloidal QDs (PbS/CQDs) as inks. Bakulin *et al.* (2013) explained the early trapping dynamics that strongly depend on the nature of the ligands used for QD passivation. Yuan *et al.* (2016) worked to improve the photocurrent in QDSSCs by employing $Pb_xCd_{1-x}S$ alloy QDs as photosensitizers and achieved an efficiency of 3.2 %. Du *et al.* (2016a) reported Zn-Cu-In-Se QDSSCs with an efficiency of 11.6 %.

Jiao *et al.* (2015) studied on band gap engineering in core shell structure of CdTe/CdSe QDs where the cell performance is limited by low photovoltage. In order to improve photovoltage and thus to improve the efficiency of the QDSSC, core shell QD with much widened conduction band offset is analysed by substituting Cd with Zn and forming ZnTe/CdSe core shell QDs . Thus, photocurrent seem to be improved along with enhanced QDSSC efficiency. High charge recombination with low hole transfer rate is one of limiting parameter reducing maximum power deliverance of CdSe QDs based QDSSC. Bang *et al.* (2011) reported CdSe with fullerene nanocomposite for supressing recombination in CdSe based QDSSC. Kim *et al.* (2015a) enhanced the efficiency by suppressing recombination in CuInSe based QDSSC by controlling ZnS overlayer thickness. Both interfacial recombination at the electrolyte and nonradiative recombination occurring within QDs are significantly suppressed with ZnS coating and thus overall QDSSC performance seem to be enhanced. Rao *et al.* (2015) studied the effect of doping of Mn on CdTe/ZnS based QDSSC cells. Outer inorganic ZnS shell for CdTe QDs along with Mn_{2+} doping resulted in enhanced photochemical stability and photoluminescence of CdTe QDs and thus improving QDSSC performance.

Tatsuma *et al.* (2015) reported on importance of QD layer thickness and strategies to make effective use of light by effectively placing QDs at optimum distance from the FTO electrode. It is suggested to place QDs preferably at a far distance from FTO for improved QDSSC performance. Tatsuma *et al* 2015 also studied the importance in dopant concentration with QDs in QDSSC. Their study showed variation in QDSSC performance with the introduction of plasmonic Ag nanocubes as dopant to colloidal PbS QDs for ZnO nanowire based QDSSCs. Initially an improved photocurrent value and power conversion efficiency was observed but with excess of Ag nanocubes resulted high charge recombination and thus lowering of photocurrent and efficiency. Supporting to above work Muthalif *et al.* (2017) also studied the effect of Cu-doping in CdS QDs for QDSSC. The study also confirmed that with optimum dopant concentration charge recombination rate is suppressed with increased electron lifetime while high level doping concentration greatly increased the charge recombination rate and lowers the QDSSC cell performance. Ardalan *et al.* (2011) reported on self assembled monolayer (SAM) CdS based QD and showed tripled enhanced efficiency for QDSSC with SAM CdS due to the presence of a recombination barrier with SAM utilized CdS QDSSC. Shibu *et al.* (2012) studied on QDs capped with fullerene for solar light harvesting. Fullerene-shell acts as electron acceptor and also a physical protection layer suppressing photocorrosion of QDs.

From the above literatures it can be inferred that to achieve a high efficiency in QDSSC several parameters have to be scientifically investigated. With proper band engineering of QDs much wider absorption range along with fast charge separation can be achieved. Still proper band tuning of QDs is a challenging issue. Recombination losses occurring at various QDSSC interfaces is another challenging problem that need to be suppressed for improved cell performance. Optimum QD loading concentration is also essential for improved QDSSC performance. Suppression of photo corrosion of QD is another major problem related to QDSSC. Though addition of dopant with QDs enhances QDSSC performance, the determination of optimum dopant concentrations is another challenging issue. Unfortunately, from these recent studies it can be inferred that these challenging issues are still remaining and have to be controlled for improving the performance of QDSSCs. Therefore, to achieve this goal, a number of areas of QDSSC have to be further explored.

SCOPE OF THE THESIS

The scope of the current thesis work is to develop TiO_2 based photoanode with controlled morphology, synthesis of novel QD materials as sensitizers and modify the standard electrolyte and CE to improve the overall performance of QDSSC.

The main objectives of the thesis are as follows;

> To prepare photoanode by preparing TiO_2 nanostructures and to study the effect of morphology of TiO_2 on the performance as a photoanode for QDSSC.

> To synthesis CdS, InSb, GQDs by optimizing the synthesis conditions and to analyze size, morphology and optical properties of the QDs by XRD, UV-Vis, SEM and TEM analysis.

> To prepare the QDs loaded photoanode and study their structural and morphological properties.

> To fabricate QDSSC and study the performance by modifying the sensitizer using novel green CdS QDs, InSb QDs and GQDs.

> To improve the photovoltaic performance of CdS based QDSSC by modifying standard polysulphide electrolyte with additives and CE.

CHAPTER 2

SYNTHESIS AND CHARACTERIZATION OF TiO₂ NANOSTRUCTURES WITH CONTROLLED MORPHOLOGY FOR QUANTUM DOT SENSITIZED SOLAR CELL APPLICATIONS

INTRODUCTION

Titanium dioxide (TiO_2) is a well known semiconductor material and widely used for various applications due to its excellent physicochemical properties (Kwoka *et al.* 2017; Li *et al.* 2011;Tatsuma *et al.* 2011). TiO_2 has wide band gap (3.37 eV) and large exciton binding energy (60 meV) (Chen *et al.* 2007). The significant variations in its light absorption and emission at nano scale make it as a material of interest for various applications. Hence, TiO_2 nanostructures find applications in many fields such as photocatalysis(Yu *et al.*2007), solid state photovoltaics (Harris *et al.* 1992), ultra violet light emitters (Hou *et al.* 2015), band pass filters (Hayashi *et al.* 2017), water purification (Guayaquil-Sosa *et al.* 2017) photo catalytic de-composition (Wu *et al.* 2011), self-cleaning glass (Kim *et al.* 2014), pigmentary industry (Gratzel *et al.* 1994), dye sensitized solar cells (Yan *et al.* 2017), rechargeable batteries (Du *et al.* 2001) and so on. One dimensional nanostructure of TiO_2 has been the focus of research in recent decades particularly in the form of nanotubes because of its exciting features such as directionality, higher surface area and its functionality (Zhao *et al.* 2002; Zhu *et al.* 2001).

Moreover, TiO_2 is one of most promising materials that can be used as a photoanode material for sensitized solar cell applications (Min *et al.* 2010). In sensitized solar cell, photoanode plays an important role to provide large specific surface area for adsorption of sensitizer molecules (Mor *et al.* 2006). Apart from that, it also helps in charge diffusion pathway for transport of injected carriers to external circuitry (Fang *et al.* 2011). Further the morphology, structure, crystallinity and surface state of nanomaterial used as photoanode can significantly influence the light harvesting and overall energy conversion effciency of solar cells (Seo *et al.* 2008). The optimization and control of synthesis parameters are essential for the synthesis of nanostructured materials (Zarate *et al.* 2007;Du *et al.* 2016b).

Number of methods were reported for preparation of TiO_2 nanostructures including template assisted method, hydrothermal method, and chemical treatments of fine TiO_2 particles and electrochemical anodic oxidation of pure titanium sheet. Kasuga *et al.*(1998) reported the synthesis of TiO_2 by using NaOH and TiO_2 and studied its properties. Chen *et al*(2002). studied the formation of TiO_2 using different bases such as KOH, LiOH and NaOH, but tubular structure was formed only when NaOH used as precursor . Peng *et al.* (2010) used microwave assisted hydrothermal treatment for the synthesis of TiO_2, at constant temperature (135 °C) and different irradiation times. Fang et al reported that TiO_2 nanotubes were obtained by immersing membrane into the sol for a short period whereas solid TiO_2 fibrils were obtained after immersion of membrane for longer times (Fang *et al.*2011). Yoriya *et al.*(2008) and co-workers synthesized the uniform and highly ordered TiO_2 nanotube arrays by anodic oxidation of a pure titanium sheet in a hydrofluoric acid (HF) based aqueous electrolyte. Kuo *et al.* (2009) studied the effect of pH on synthesis of TiO_2 nanotubes.

Titaniumtetrachloride, titaniumtetraisopropoxide, titanium butoxide and P25 are widely used as precursors for the synthesis of TiO_2 nanostructures. However, these source materials are comparatively expensive and toxic in nature. Therefore, it is highly essential to find a low cost and environmentally friendly source material to synthesis TiO_2 nanostructures especially for large scale applications. In this chapter, low cost, nontoxic and commercially available polycrystalline TiO_2 is used as a source material for synthesis of TiO_2 nanostructures. Structural and morphological variations of a photoanode material greatly influence the overall performance of solar cell (Sarkar *et al.* 2016a). Therefore, synthesis of TiO_2 with controlled morphology is highly imperative to study the effect of morphology on electron transport of photoanode (Hu *et al.* 2011; Chen *et al* .2002). Although the effect of the morphology of TiO_2 nanostructures on the electron transport were reported further systematic investigation will be highly useful to understand the effect more in detail. Therefore, in this chapter, the TiO_2 nanostructures were synthesized by two methods with different morphologies and effect of morphology on optical and electrical properties of TiO_2 nanostructures have been investigated.

EXPERIMENTAL

2.2.1 Synthesis of Titanium dioxide Nanostructures

Polycrystalline TiO_2 (80 % anatase, 20 % rutile (Merck), sodium hydroxide (NaOH), hydrochloric acid (HCl) (Alfa aesar) were used as source materials without further purification. In order to study the effect of synthesis parameters on TiO_2 nanotube formation, the material was synthesized by sol gel and hydrothermal methods. Further to study the effect of growth time on the formation of TiO_2 nanostructures under hydrothermal condition, hydrothermal reaction was carried out at different time period such as 12, 24 and 48 h.

For sol-gel synthesis, 2 g of polycrystalline TiO_2 powder was mixed with 10 M of NaOH and stirred for 24 h with addition of 0.1 M HCl. The obtained white color precipitate was washed and centrifuged at 5000 rpm until pH becomes neutral. The collected samples were separated into three parts. One part of the sample was freeze dried at −80 °C, second part was dried at 100 °C and the third part was calcined at 450 °C. In a typical hydrothermal synthesis, 2 g of polycrystalline TiO_2 powder was added to 10 M of NaOH aqueous solution and stirred for 2 h. The mixture was transferred into a teflon lined vessel and loaded on to an autoclave and kept in furnace at 180 °C. Three experiments were conducted at different reaction periods of 12, 24 and 48 h.

Table 2.1 Details of the synthesized samples

Sample Codes	Synthesis Method	Time	Sample Processing Temperature
D1	Sol gel method	24 h	Freeze dried at -80 $^\circ$C
D2			Dried at 100 $^\circ$C Calcined at 450 $^\circ$C
D3			
E1	Hydrothermal (180 $^\circ$C) method	12 h.	Freeze dried at -80 $^\circ$C Dried at 100 $^\circ$C
E2			
E3			Calcined at 450 $^\circ$C
F1	Hydrothermal (180 $^\circ$C) method	24 h	Freeze dried at -80 $^\circ$C Dried at 100 $^\circ$C Calcined at 450 $^\circ$C
F2			
F3			
G1	Hydrothermal (180 $^\circ$C) method	48 h	Freeze dried at -80 $^\circ$C
G2			Dried at 100 $^\circ$C Calcined at 450 $^\circ$C
G3			

After hydrothermal reaction 0.1 M of HCl was added to neutralize the pH and resulted products were washed three times using double distilled water. The resulted white product was isolated by centrifugation at 6000 rpm and transferred to petridish. After that the sample was treated at different conditions such as freeze dried at $-80\,°C$, dried at $100\,°C$, and calcined at

$450\,°C$. The prepared samples were named as D1, D2, D3 for sol gel synthesized samples, E1, E2, E3 for hydrothermally prepared 12 h samples, F1, F2, F3 for hydrothermally prepared 24 h samples and G1, G2, G3 for hydrothermally prepared 48 h samples as shown in Table 2.1.

CHARACTERIZATION OF TiO₂ NANOSTRUCTURES

The structural properties of TiO_2 obtained by sol-gel and hydrothermal methods were examined by X-ray diffraction (XRD) analysis using X-ray diffractometer (Bruker D8 advance diffractometer) instrument with Cu Kα 1.54 Å radiation. The XRD patterns were recorded in the range of 2Θ from 20 to 60 ° with a stepwise increment of 0.02 and a count time of 5 s. Morphology of TiO_2 samples were investigated by scanning electron microscopy (SEM) using TESCAN VEGA3 system. The UV-vis diffuse reflectance spectra (UV-DRS) of TiO_2 samples were recorded using UV-Vis spectrophotometer (UV-2450 Shimadzu spectrophotometer). Jasco FP-8600 spectrofluorometer was used to study the photoluminescence properties of the samples. Thermal stability of the samples was studied by thermalgravimetric analysis (TG/DTA 6300). The presence of functional groups in the samples were studied by Raman (Peak seeker 532) and FTIR (Jasco 6600) spectrophotometers. XPS analysis of prepared samples was carried out by SHIMADZU ESCA 3400 X-ray photoelectron spectrophotometer. Electrical properties of the TiO_2 samples were measured by Hall effect measurement system (Model H5000 MMR technologies, USA). The prepared samples were made into pellets by high pressure and temperature sintering process. Pellets of

1.5 mm thick and 16 mm diameter were prepared by applying pressure of 200 mbar and sintered at 450 °C for electrical measurements. Before the measurement, ohmic contacts were made by deposition of silver paste at four corners of the pellets. The measured resistivity is the bulk resistance as the thickness of the pellets was considered for the measurement.

RESULT AND DISCUSSION

Structural and Morphological Analysis of TiO_2 Nanostructures

Figure 2.1-2.4 shows the XRD patterns of TiO_2 nanostructures synthesized by sol gel (sample D) and hydrothermal methods at reaction period of 12 (sample E), 24(sample F) and 48(sample G) h. The XRD patterns of sol gel synthesized samples are shown in Figure 2.1. The results confirm the anatase phase of TiO_2 nanostructures. Moreover, the sharp and intense diffraction peaks with small full width at half maximum (FWHM) illustrated the high crystalline nature of sol gel synthesized samples. The intensity of peaks was relatively high for calcined sample compared to freeze dried sample due to the crystallization at high temperature. The high crystallinity of the sol-gel synthesized sample is mainly attributed due to the growth of particles in the solgel medium under ambient temperature and pressure.

Figure 2.2-2.4 shows the de-convoluted XRD patterns of hydrothermally prepared TiO_2 nanostructures at different reaction periods of 12, 24 and 48 h. On deconvolution of diffraction peaks, the XRD patterns clearly show the mixed phases of anatase and rutile TiO_2 in agreement with the standard diffraction data (JCPDS 21–1272 and 21–1276). Most of the diffraction peaks were matched with anatase phase of TiO_2 and a few peaks found at 2Θ of 27.4° and 35.6° corresponds to (110) and (101) planes of rutile phases (Figure 2.2-2.4). In addition, a few broad peaks of $Ti(OH)_2$ were also observed in the hydrothermally prepared samples named as T in XRD patterns. All the peaks were indexed in the deconvoluted XRD patterns. Peak shift was observed due to strain resulting from the mixed phases and other defects in the hydrothermally prepared samples. Moreover, the XRD patterns of hydrothermally prepared samples show low intensity peaks with large FWHM which indicate the low order crystallinity of the samples.

The diffraction patterns and peak positions of TiO_2 nanostructures were well matched with tetragonal structure of TiO_2 with anatase phase (JCPDS-21-1272). In addition, the cell parameters of a = b = 0.377 nm, c = 0.943 nm and $\alpha = \beta = \gamma = 90°$ were in line with the standard data (JCPDS- 21-1272) (Nakahira *et al.* 2010, Chen *et al.* 2009, Schulte *et al.* 2010, Filippo *et al.* 2015; Xiao *et al.* 2010).

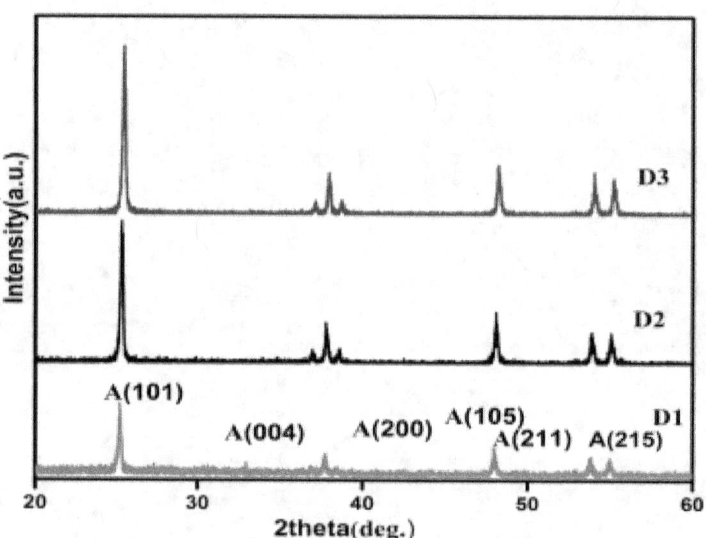

Figure 2.1 XRD patterns of sol gel synthesized TiO₂ (D1, D2, D3)

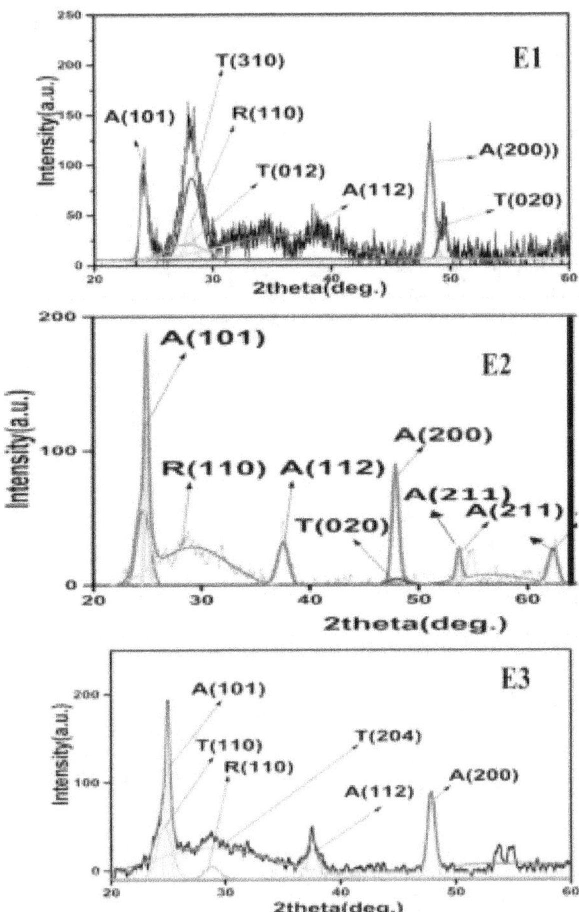

Figure 2.2 Deconvoluted XRD patterns of hydrothermally prepared 12 h samples (E1, E2, E3) 1-Freeze dried at -80 °C,2-dried at 100 °C, 3-Calcined at 450 °C

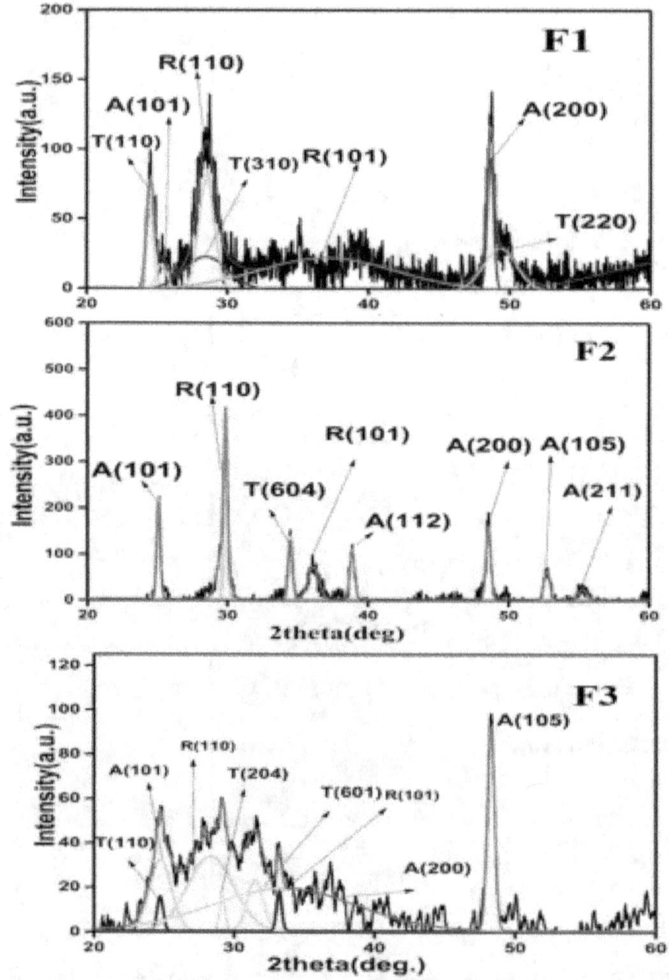

Figure 2.3 Deconvoluted XRD patterns of hydrothermally prepared 24 h samples (F1, F2, F3) 1-Freeze dried at -80 °C,2-dried at 100

°C,3-Calcined at 450 °C

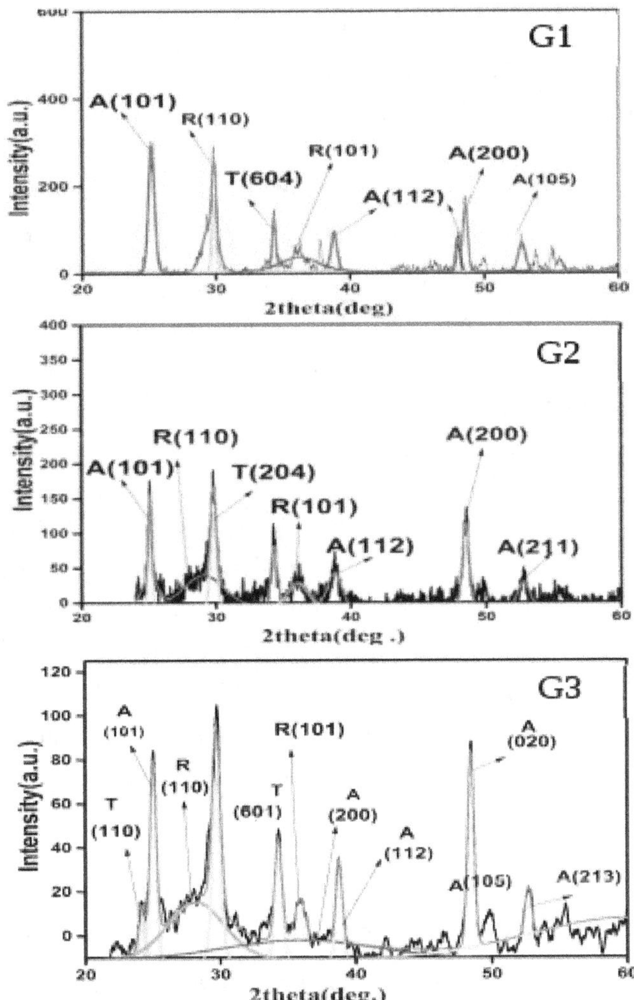

Figure 2.4 Deconvoluted XRD patterns of hydrothermally prepared 48 h samples (G1, G2, G3) 1-Freeze dried at -80 ∘C,2-dried at 100 ∘C,3-Calcined at 450 ∘C

The morphology of TiO_2 nanostructures was examined by SEM analysis. Figure. 2.5 (a–d) show SEM images of TiO_2 nanostructures prepared by sol gel and hydrothermal methods. The SEM images confirm the spherical morphology of the sol-gel synthesized TiO_2 nanostructures in Figure. 2.5 a (i–iii). Further, the particles of samples D1, D2, D3 were monodispersed as can be seen from the SEM images.

Figure. 2.5 b (i, ii, iii) show the SEM images of hydrothermally prepared TiO_2 at the reaction period of 12 h (E1, E2, E3). The sheet like structures were clearly observed in the freeze dried and 100 ᵒC dried samples in Figure 2.5 b (i) and (ii). While spikes were observed in the calcined samples in Figure 2.5 b (iii). The SEM images of hydrothermally prepared samples (F1, F2, F3, G1, G2, G3) with relatively longer reaction period Figure 2.5 (c) and

(d) clearly shows the tube like structures of TiO_2 with the diameter in the range of 50 to 100 nm and length of several microns. The sheet like structure was scrolled to form tubular morphology during the longer reaction period under hydrothermal conditions (24 and 48 h) (Chen *et al.* 2009, Filippo *et al.* 2015; Li *et al.* 2009). Moreover, the number of tubes were increased in the freeze dried (F1) and 100 ᵒC dried (G1) samples compared to calcined sample. Further it was observed that the tube diameter increased with reaction time Figure 2.5 c (i) & d (i, ii).

The observations of the microstructural change in TiO_2 under hydrothermal condition suggested that nanotube structures were formed from sheet like microstructure in Figure. 2.5 b (i) with increase in the reaction period as the sheets exfoliate to form tube structure. The SEM images showed collapse of tubular morphology in samples calcined at high temperature Figure 2.5 b (iii), 2.5 c (iii), 2.5 d (iii) due to the removal of -OH groups from the surface of the tubes.

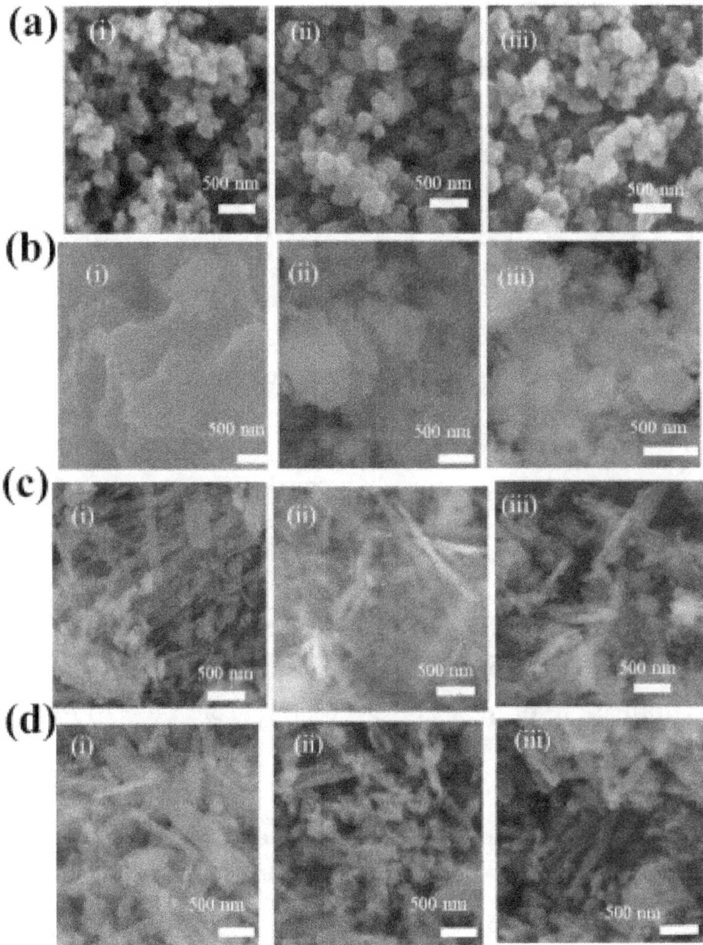

Figure 2.5 SEM images of TiO$_2$ samples (a) solgel synthesized (i) D1 (ii) D2 (iii) D3. (b)Hydrothermally synthesized 12 h(i) E1 (ii) E2

(iii) E3 (c) Hydrothermally synthesized 24 h (i) (F1) (ii) F2 (iii) F3. (d) Hydrothermally synthesized 48 h (i) G1 (ii) G2 (iii) G3

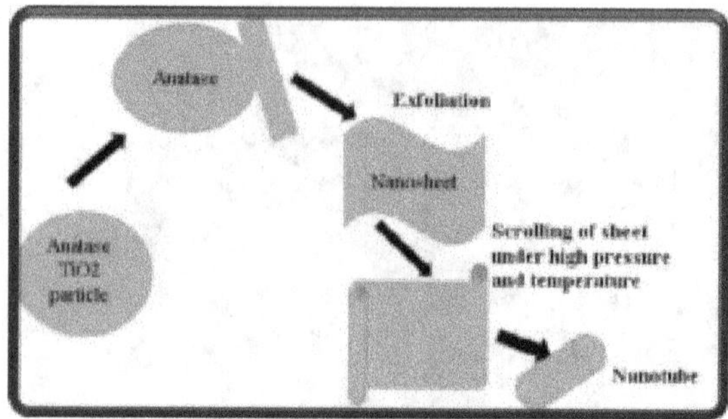

Figure 2.6 Schematic diagram for formation mechanism of TiO₂ nanotubes

The detailed mechanism of formation of TiO₂ nanotube is explained as follows; When polycrystalline TiO₂ raw material was made to react with aqueous NaOH, it initiates the process of breaking Ti-O-Ti bonds for the formation of Ti-O-Na and Ti-OH. Ti-O-Na and Ti-OH bonds get reacted with HCl and double distilled water resulting formation of new Ti-O-Ti bonds or Ti- O-H-O-Ti hydrogen bonds. The decrement in bond distance of Ti atoms on surface results in folding of sheets to form tubular structure. The exfoliation of nanosheets occurs only in the presence of temperature and pressure hence, the tubular structure was formed only under hydrothermal conditions. The formation mechanism of TiO₂ nanotube is schematically shown in Figure 2.6.

FTIR Studies of TiO₂ Nanostructures

FTIR spectra were recorded for TiO₂ nanostructures prepared by sol gel and hydrothermal methods (48 h) (Chen *et al.* 2009; Li *et al.* 2011). The FTIR spectra showed in Figure 2.7 and the three IR absorption bands centered at 3400 cm⁻₁, 1630 cm⁻₁ and 950 cm⁻₁ Figure 2.7 (a) and (b) for TiO₂ nanostructures prepared by sol gel method and hydrothermal (48 h) methods. The broad peak at 3400 cm$_{-1}$ Figure 2.7 (b) was due to the O-H asymmetrical stretching of physically absorbed water and the peaks at 1630 cm$_{-1}$ and 950 cm–1 were due to H-O-H bending. Further the FTIR results indicated that most of Na+ replacement occurred with H+ in TiO_2 nanotubes during acid washing. The bands observed at 720–970 cm–1 in Figure 2.7 (b) in TiO_2 nanostructures prepared by hydrothermal method (48 h) were the characteristic peaks of TiO_2 nanotubes which were absent in sol gel synthesized samples. The FTIR spectra shows a peak related to Ti-O-Ti frameworks which were responsible for the tube formation and theses peaks were not observed in the sol gel prepared samples.

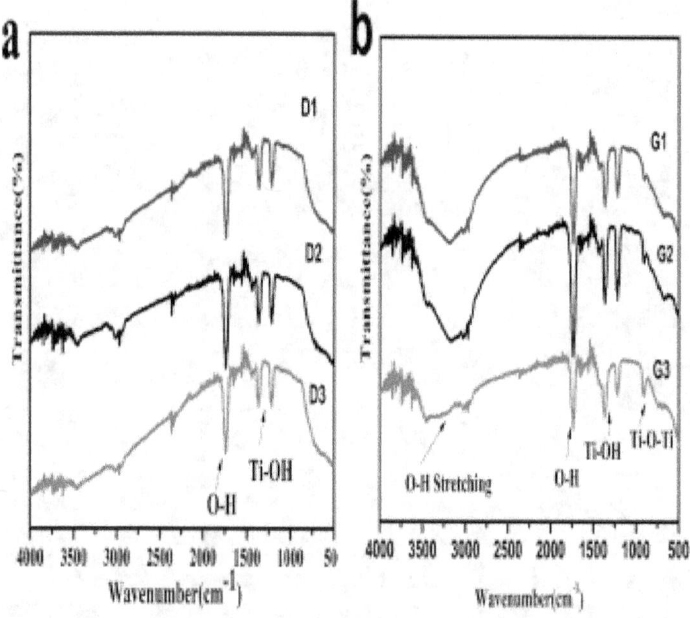

Figure 2.7 FTIR Spectra of TiO₂ samples (a) sol gel synthesized TiO₂ (D1, D2, D3); (b) hydrothermally prepared 48 h samples (G1, G2, G3), 1-freeze dried at -80 ◦C, 2-dried at 100 ◦C, 3-calcined at 450 ◦C

Raman Analysis of TiO₂ Nanostructures

Raman spectra of prepared TiO_2 samples were recorded and shown in Figure 2.8 (a) and (b). The Raman spectra of sol gel synthesized samples show three distinct peaks at 379, 415 and 640 cm-₁ Figure 2.8 (a) corresponds to anatase phase of TiO_2 (Bellat *et al.* 2015). The Raman spectra of hydrothermally prepared samples show the peaks at 279 and 373 cm-₁ corresponds to stretching vibration of Ti-O-Na bonds and the peak at 674 cm-₁ was assigned to Ti-O-Ti stretching vibrations of TiO_6 octahedral (Kiatkittipong *et al.* 2010; Grigorieva *et al.* 2010). The presence of peaks at 279, 373 425 and 610 cm-₁ were due to rutile phase of TiO_2 which were absent in the sol gel synthesized samples Figure 2.8 (b). The peaks at 373 and 674 cm-₁ were due to the presence of tubular nanostructure in the hydrothermally prepared samples. The peak at 786 cm-₁ in Figure 2.8 (b) corresponds to Ti-O-H stretching vibration of titanium hydroxide in TiO_2 nanotubes which was already confirmed by XRD in Figures. 2.2-2.4.

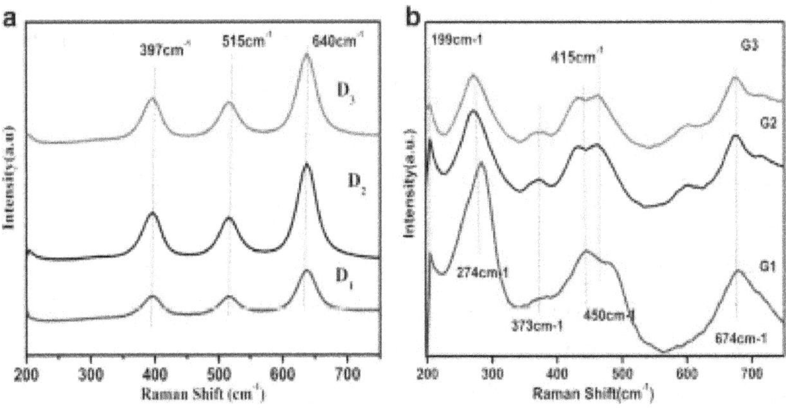

Figure 2.8 Raman spectra of (a) sample (D1, D2, D3) (b) sample (G1, G2, G3)

Thermogravimetric Analysis (TGA) of TiO₂ Nanostructures

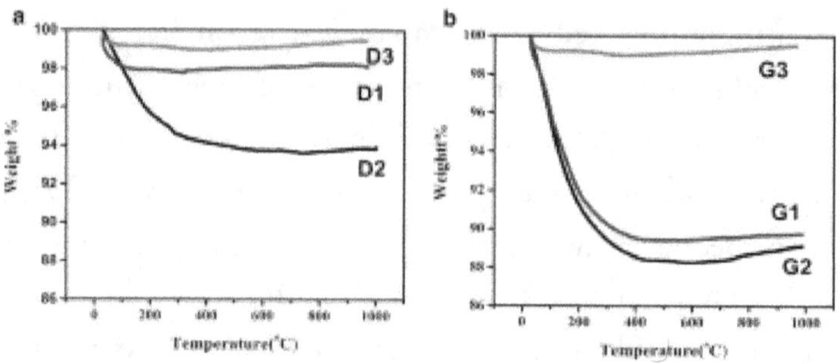

Figure 2.9 TGA of TiO₂ samples (a) Solgel synthesized (D1, D2, D3) (b) Hydrothermally synthesized (G1, G2, G3)

In order to study the thermal stability of TiO₂ nanostructures, TGA was performed for both sol gel and hydrothermally prepared samples. TGA result indicated a weight loss related to evaporation of absorbed water below temperature of 150–200 ∘C. The dehydration of samples was observed at 350– 460 ∘C. The total weight loss of sol gel synthesised TiO₂ nanoparticle was about 6 % and 3 % for freeze dried (D₁) and dried at 100 ∘C (D₂) respectively. The calcined (D₃) sample seem to be stable with less than 1 % of weight loss as shown in Figure. 2.9 (a). The total weight loss for TiO₂ nanotube synthesised by hydrothermal method was about 12 % Figure 2.9 (b). The difference in water loss peculiarities could be interpreted by relating with crystallographic and morphological studies. The hydrothermally prepared TiO₂ nanotube showed higher water loss especially for the freeze dried and 100 ∘C dried samples (12

%) Figure 2.9 (b). The TGA study confirmed that more amount of O-H bonding present in -80 ∘C freeze dried and 100 ∘C dried samples were lost compared to calcined samples (Hossain *et al.* 2010) which resulted higher weight loss. Further, the TGA result indicated that the calcined samples prepared by both methods exhibited good thermal stability.

UV VIS DRS studies of TiO₂ Nanostructures

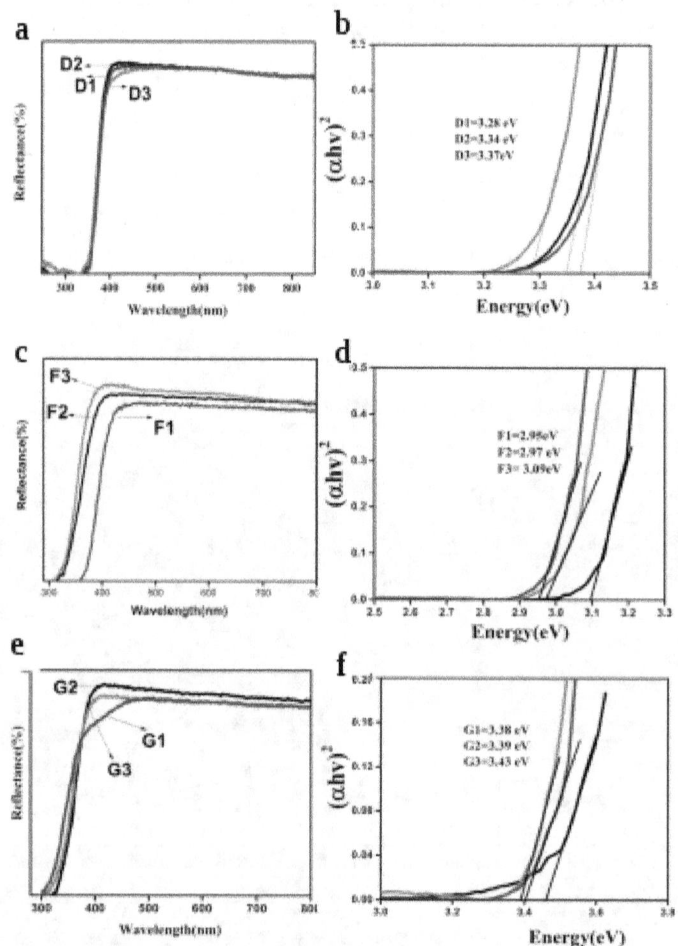

Figure 2.10 UV-Vis absorption spectra and Tauc plot of TiO₂ samples (a and b) sol gel D1, D2, D3 (c and d) hydrothermally prepared 24 h F1, F2, F3 (e and f) hydrothermally prepared 48 h (G1, G2, G3)

Figure 2.10 (a–e) shows the UV-Vis diffuse reflectance spectra (DRS) and Tauc plot of sol gel (D) and hydrothermally prepared (24 (F), 48

(G) h) TiO_2 samples. As shown in Figure 2.10 (a) cut off wavelength was about the same for all the sol gel synthesized samples. Band gap energy was calculated from Tauc plot and by extrapolating the tangent to the x-axis. The calculated band gap energies of sol gel derived samples (D) were in the range of 3.28 to 3.37 eV shown in Figure 2.10 (b). The DRS spectra of hydrothermally prepared samples in Figure 2.10 (c) show clear variation in the absorption edges for calcined (F3) and 100 °C dried (F2) samples compared to freeze dried sample (F1) in Figure 2.10 (c). The calculated band gap energies of F1, F2 and F3 were 2.95, 2.97 and 3.09 eV, respectively Figure 2.10 (d). The observed reasonable variation in the bandgap of TiO_2 was mainly attributed to the mixed phases of anatase and rutile of TiO_2 in the sample as confirmed by XRD and Raman analysis. Similarly, a reasonable variation in the absorption edges thereby the bandgap of the hydrothermally prepared 48 h samples (G1, G2 and G3) were clearly observed in Figure 2.10 (f). The presence of anatase (3.2 eV) and rutile (3 eV) phases causes bandgap shift from 2.95 eV to 3.48 eV for hydrothermal samples. The optical studies revealed that the performance can be enhanced by tuning the band gap of the TiO_2 material. It is emphasized that the controlling of synthesis parameters for TiO_2 material is rendered as significant factor for tuning the bandgap thereby increasing the performance.

Photoluminescence study of TiO_2 Nanostructures

Figure 2.11 (a-c) shows PL spectra of sol gel (D) and hydrothermally prepared 24 h (F) and 48 h (G) TiO_2 samples. The two emission peaks were observed at 365 and 385 nm, respectively for all samples. It can be inferred that the peak at 385 nm is prominent for sol gel synthesized samples which contains the pure phase of anatase, while two peaks at 365 and 385 nm were prominent for hydrothermally prepared samples. The prominent two peaks in hydrothermally prepared samples were due to the presence of mixed phases of anatase and rutile as observed in the XRD and Raman analysis. The emission at two different wavelengths was due to difference in band gap of anatase and rutile phases of TiO_2 (Bavykin *et al.* 2005, Ma *et al.* 2017, Chetibi *et al.* 2017; Lai *et al.* 2010). The PL study demonstrated that the emission performance can be enhanced by controlling synthesis parameters and thus recombination of surface trapped electrons can be relatively suppressed. The PL peaks of TiO_2 nanotubes were relatively boarder, which may be due to the alignment of rutile and anatase phase (Al-Abdullah *et al.* 2010; Kim *et al.* 2014). The band alignment favours the electron transport from anatase to rutile inducing reduced charge recombination and thus TiO_2 nanotube with mixed phase is a promising photoanode material for solar cell applications.

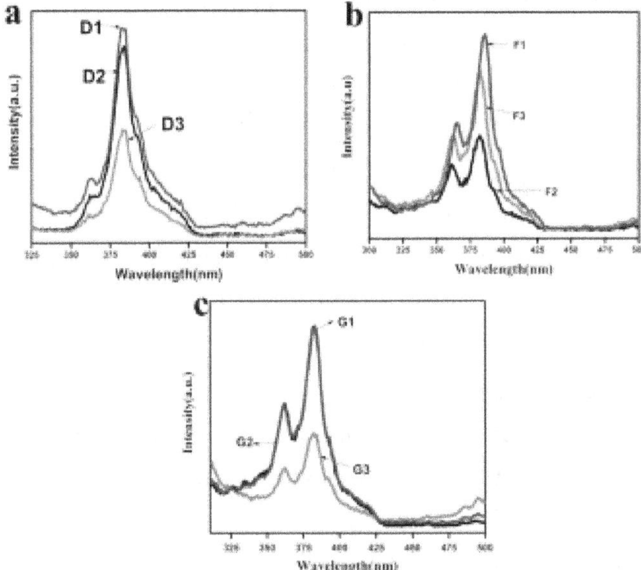

Figure 2.11 Photoluminescence spectra of TiO_2 Samples (a) solgel synthesized (D1, D2, D3 (b) hydrothermally prepared 24 h (F1, F2, F3) (c) hydrothermally prepared 48 h (G1, G2, G3)

XPS Analysis of TiO₂ Nanostructures

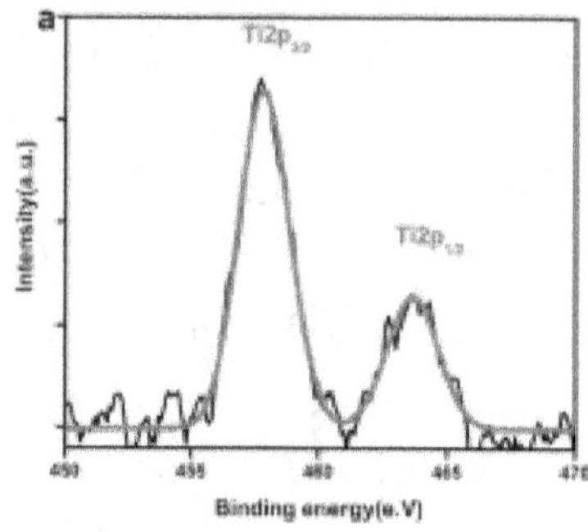
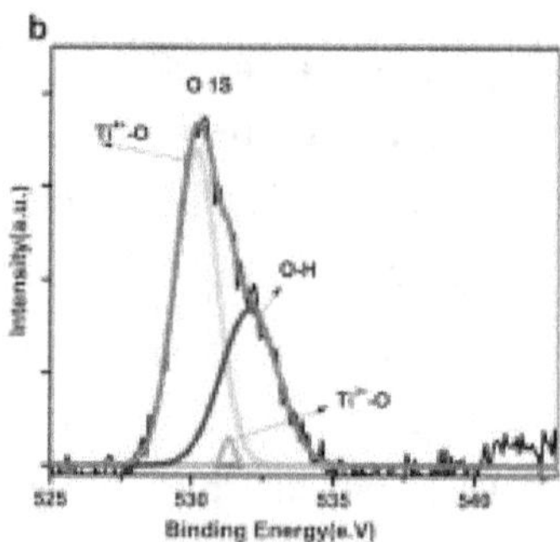

Figure 2.12 XPS spectra (a) Ti 2p (b) O1s of hydrothermally prepared TiO₂ nanostructure (sample F2)

The XPS spectra of hydrothermally prepared 48 h samples (G) are shown in Figure 2.12 (a) and (b). The core level spectrum of Ti Figure 2.12 (a) shows the doublet peaks of Ti 2p$_{3/2}$ and Ti 2p$_{1/2}$ at 458.88 and 464.01 eV, respectively. The doublet peaks were separated with the binding energy of 7 eV which arises from TiO₂ spin orbit-splitting. These peaks were consistent with Ti$_{4+}$ in TiO₂ lattice (Hossain *et al.* 2010). Figure 2.12 (b) shows the core level spectrum of oxygen 1s of TiO₂ prepared by hydrothermal method. The deconvoluted O 1s spectrum shows two peaks at 530.31 and 532.02 eV, respectively. The peak at 530.31 eV corresponds to lattice oxygen and the shoulder peak at 532.02 eV represents the surface oxygen. The deconvolution of the O 1s spectrum revealed the binding energies of the individual components to be 530.31 (Ti$_{4+}$–O), 531.1 (Ti$_{3+}$–O), and 532.36 eV (OH-), in agreement with previous work on various titanium oxide phases (Guo *et al.* 2009, Bharti *et al.* 2016, Othman *et al.* 2014; Han *et al.* 2013). XPS data showed weak binding for Ti$_{3+}$- O atom compared to Ti$_{4+}$-O atom as the intensity of Ti$_{3+}$–O related peak is relatively low. XPS data shows some difference in bonding of Ti-O as explained in nanotube formation mechanism in comparison with SEM and Raman analysis.

Electrical Studies of TiO_2 Nanostructures

Electrical resistance, resistivity, conductivity and mobility of 100 °C dried TiO_2 nanoparticles and nanotubes are shown in Table 2.2 and compared with reported data. Hall effect measurement was carried out at temperature of 300 K. It can be seen that hydrothermally prepared TiO_2 nanotubes for 24 h showed relatively higher electrical characteristics such as high mobility of 7.05 (m_2/V. s) and conductivity of 316.4 × 10-4 ($D._{-1}$ m-1) compared to the TiO_2 nanoparticles and nanotubes (48 h). Further, the measured value of electrical conductivity of 24 h sample was relatively higher than the previously reported values (Hershah *et al.* 2012). From the results it is obvious that TiO_2 nanotubes prepared at 24 h exhibit low resistivity and high conductivity. Therefore, it can be a suitable candidate for photoanode material with low resistivity for ease electron transport in QDSSCs. As a consequence, the synthesis condition of 24 h reaction period in hydrothermal method was observed to be an optimum condition for synthesis of TiO_2 nanotubes as photoanode material.

Table 2.2 Measured resistance, resistivity, conductivity and mobility of D2, F2 and G2 samples

Sample Name	Resistance (MD.)	Resistivity (D.m)	Conductivity (D._{-1} m-1)	Mobility (m2/V. s)
TiO_2 nanoparticles sample D2	0.496	4.20 X10-3	2.38 X 10-4	1.41
TiO_2 nanotubes (24 h) sample F2	0.047	0.0316 X10_3	316.4 X 10-4	7.05
TiO_2 nanotnbes (48 h) sample G2	0.771	0.27 X 10_3	37.03 X 10-4	5.31
TiO_2 nanowires (Alamelu *et al.* 2018)	0.26	4.47 x 10_3	2.24 X 10-4	

CONCLUSION

TiO_2 nanostructures were synthesized by sol gel and hydrothermal methods. The structural and morphological properties of the prepared materials were studied by XRD and SEM analysis. The morphological evolution under hydrothermal condition was clearly observed from the SEM images. The formation of nanotubes under hydrothermal conditions was further confirmed by Raman and FTIR analysis. The PL study shows the emission peaks at two different wavelengths due to mixed phase in the hydrothermally prepared samples. The electrical conductivity of the TiO_2 nanotubes prepared at 24 h was relatively higher than that of sample prepared at 48 h and reported values of similar material. The present study explored the influence of synthesis conditions on morphology, optical and electrical characteristics of TiO_2 nanostructures. Further, the experimental results demonstrated that the morphology of nanostructure has great influence in its optical and electrical properties. The low resistivity of the hydrothermal prepared TiO_2 confirms that hydrothermal synthesis as an effective method for synthesis of TiO_2 as a suitable photoanode material for sensitized solar cells.

CHAPTER 3

SYNTHESIS AND CHARACTERIZATION OF GREEN CdS QUANTUM DOTS AND APPLICATION AS SENSITIZER IN QUANTUM DOT SENSITIZED SOLAR CELLS

INTRODUCTION

QDSSCs are attracting enormous attention and thereby number of relevant research articles increasing annually (Snaith *et al.* 2014; Lu *et al.* 2011; Yang *et al.* 2017). Recently, Della *et al.* (2015) reported highest photoconversion efficiency of QDSSC of 13.43%. Different strategies have been tried in improving performance and stability of QDs in QDSSCs (Azimi *et al.* 2014; Kojima *et al.* 2009; Prabavathy *et al.* 2017). However, only limited works are reported in controlling the toxicity of QDs in QDSSCs (Kim *et al.* 2015a; Kamat 2013). The smaller size of nanoparticles can result in the capability to enter in to the human body by inhalation, ingestion, skin penetration or injections; thereby, creating the potential to interact with intracellular structures. The toxicity of nanomaterials, specifically because of occupational exposure for a period between 5 and 13 months was studied by European Respiratory Journal (Kovacs *et al.* 2009), and the health impairments were reported. Better control of nanomaterial toxicity can be achieved by the safe design of nanoparticles using greener approach (Prasad *et al.* 2010; Bai *et al.* 2019).

Loo *et al.* (2012) explained about the synthesis of Ag nanoparticles from Camellia sinensis extract. Recently Shivaji *et al.* (2018) have green synthesized CdS QDs using tea leaf extract and studied their antimicrobial activity. Consequently, green chemistry mediated synthesis of QDs is profoundly imperative to reduce its toxicity and extend its applications (Nadeem *et al.* 2018; Pan *et al.* 2014). However, green CdS based QDSSCs have been rarely investigated for QDSSC applications.

Azadirachta Indica (A. indica) is a well known green source for stabilizing the nanoparticles and used for synthesizing Ag and Pt nanoparticles (Thirumurugan *et al.* 2016; Velusamy *et al.* 2015). Howbeit, A. indica is not yet used as a stabilizing agent to synthesis CdS QDs. Therefore, in this chapter, the CdS QDs were synthesized using A. indica extract as a stabilizing agent and QDSSCs were fabricated using green synthesized CdS QDs. QDSSCs were comparatively investigated with different cell combinations and also compared with conventional SILAR CdS based QDSSCs.

EXPERIMENTAL

Materials

Cadmium chloride ($CdCl_2$, 99.99 % purity), methanol, sodium sulfide (Na_2S, 98 % purity) were obtained from Alfa aesar. All other reagents were used in analytical grade. Deionized water used throughout experiment was obtained from ultrapure water purification system.

Preparation of Plant Extract

For green synthesis of CdS QDs, A. indica plant extract was prepared using the following procedure. A. indica plant leaves were obtained from Adyar (Tamil Nadu, India) location. First, the leaves were washed with distilled water and then dried under shade. Following that, 10 g of chopped leaves was taken, with 100 mL of methanol, and kept at 24 h in dark room at room temperature. Then the extract solution was filtered through the Whatman qualitative filter paper (grade number 1) and stored at 4 °C for further use (Nadeem *et al.* 2018; Felix *et al.* 2014).

Green Synthesis of CdS Quantum Dots

The CdS QDs synthesis was rendered in two stages. In first stage

0.02 M $CdCl_2$ was added to 50 mL of A. indica extract and kept for 3 days in the dark room at ambient temperature. In the second stage, 0.5 mL of 0.025 M Na_2S was added and incubated for 4 days to form CdS nanoparticles. The resultant final solution was yellowish in color with greenish tint. As obtained precipitate was centrifuged at 10000 rpm for 10 min and washed with deionized water for three times (Shivaji *et al.* 2018; Pan *et al.* 2014). Finally, the obtained product was lyophilized for further characterization studies.

Fabrication of Solar Cell

TiO_2 was synthesized by hydrothermal method using TTIP as precursor at 240 °C for 12 h. The obtained TiO_2 was calcined at 500 °C for anatase phase and 700 °C for rutile phase. TiO_2 photoanodes were prepared by doctor blade method with active area of 0.25 cm_2. Green CdS QDs were loaded on to TiO_2 photoanode by direct absorption method. For comparative analysis, CdS QDs were loaded on to TiO_2 photoanode using SILAR method too. TiO_2 photoanodes were immersed in a solution of 0.4 M $CdCl_2$ dissolved in 1:1 ratio of methanol and water for 45 s followed by rinsing with methanol for 30 s and then immersed in 0.4 M Na_2S solution of methanol and water (1:1) for 45 s, followed by rinsing with methanol.

The deposition temperature was kept at ambient temperature (~ 32 °C). The number of deposition cycles were fixed as 8 cycles for effective QD loading (Ha *et al.* 2014, He *et al.* 2015, Lee *et al.* 2008).

ZnS was coated on the QD loaded photoanodes as a passivation layer (2 cycles) and studied their impact on the performance of QDSSC. The ZnS passivating layer was deposited over the green CdS and SILAR CdS QDs by 2 cycles of SILAR deposition using aqueous solution containing 0.1 M zinc acetate and 0.1 M Na_2S as anionic and cationic sources respectively.

The Pt CE was fabricated by using thermal deposition method. In detail, a drop of H_2PtCl_6 solution (5 mM of chloroplatinic acid in isopropanol) was dropped onto FTO substrate, followed by heating at 400 ∘C for 15 min. SILAR approach was used to prepare CuS CE on the FTO substrate. FTO substrates were cleaned ultrasonically with acetone, ethanol, and DI water for 10 min each. The cleaned substrates were dried by N_2 gas. The CuS CE was prepared by applying 4 cycles of SILAR using cationic and anionic aqueous solution of 0.5 M Cu $(NO_3)_2$ and 0.5 M Na_2S. Finally, the electrode was dried at 120 ∘C for 10 min.

The polysulphide electrolyte was prepared by mixing 0.5 M Na_2S, 0.1 M KCl and 0.2 M sulphur in ethanol and water (7:3) which was stirred in dark for 5 h (Lee *et al.* 2008; Kim *et al.* 2017). The QDSSC was fabricated by sandwiching QD loaded photoanode and Pt/CuS CE.

The surface morphology and composition of green CdS were analyzed using field emission scanning electron microscope with energy- dispersive x-ray analysis (FESEM-F E I Quanta F E G 200). Further, the shape and size of CdS sample were interpreted by high resolution transmission electron microscopy (HRTEM; JEOL TEM 2400). The functional groups of the prepared CdS materials were elucidated by FTIR (Jasco 6600) spectrophotometer. The UV−Vis absorption was recorded using Perkin Elmer Optima 5300 DV. J-V characteristics of the QDSSC were studied by solar simulator (1 SUN Oriel Class AAA). Electrochemical impedance spectroscopy (EIS) measurements were recorded by CHI electrochemical workstation (CHI

1106C) at frequency ranging from 10 MHz to 100 KHz and impedance spectra were analyzed with EC lab software1.

RESULT AND DISCUSSION

Structural and Morphological Analyses of Green CdS

Figure 3.1 shows the XRD pattern of green CdS which were well matched with standard JCPDS data (No.10-454) and confirms the cubic structure of CdS. The peaks observed in the XRD patterns of CdS nanoparticles at 2θ of 26.4,30.2, 43.6, 51.9 and 70.4, are corresponds to (111), (200), (220),

(311) and (331) planes of the face centered cubic CdS. Among the diffraction peaks, (111) peak was having large intensity which shows the preferred orientation of the particles. Broadened diffraction peaks with high full width at half maximum were possibly due to smaller dimension of the particles.

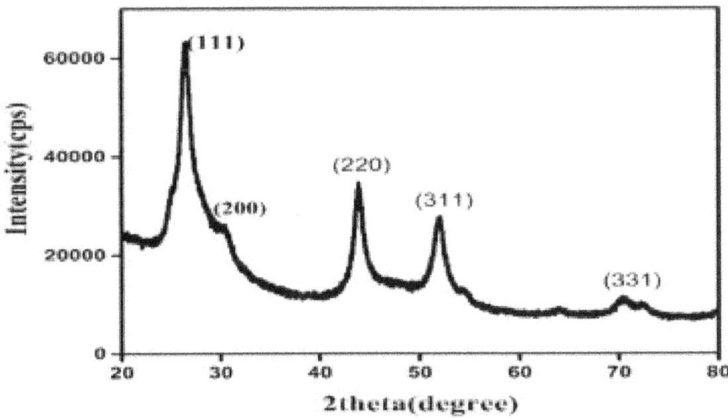

Figure 3.1 X-ray diffraction pattern of green CdS QDs

Figure 3.2 (a) shows FESEM image which confirms the spherical morphology of CdS nanoparticles and the particles were highly agglomerated. The composition of the prepared CdS was confirmed by EDAX spectrum in Figure 3.2 (b) and the composition ratio was shown as inset in Figure 3.2 (b).

In the EDAX spectrum, the strong peaks corresponding to Cd and S were clearly observed from Figure 3.2 (b). The presence of oxygen peak in EDAX spectrum showed presence of small amount of organic residuals with CdS QDs.

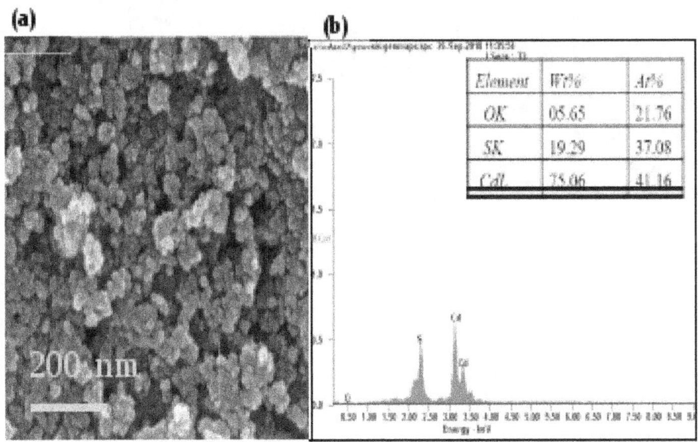

Figure 3.2 FESEM image (a) and EDAX with elemental composition (b) of green CdS QDs.

Figure 3.3 (a, b) shows the TEM and HRTEM images of green CdS nanoparticles. The images confirm the spherical morphology of CdS QDs with homogeneous size distribution of particles. The size of the particles were in the range of 2 -5 nm as can be seen from the histogram Figure 3.3 (c). From the HRTEM images, it is obvious that the size of the particles were in the range of excitonic Bohr radius of CdS (~3 nm) thereby the prepared green CdS can be considered as QDs. The inset of Figure.3.3 (b) shows the HRTEM of a single QD with clear lattice fringes and the inter planar distance of 0.33 nm is observed, which corresponds to preferred orientation of (111) of CdS QDs, as confirmed from XRD pattern from Figure 3.1.

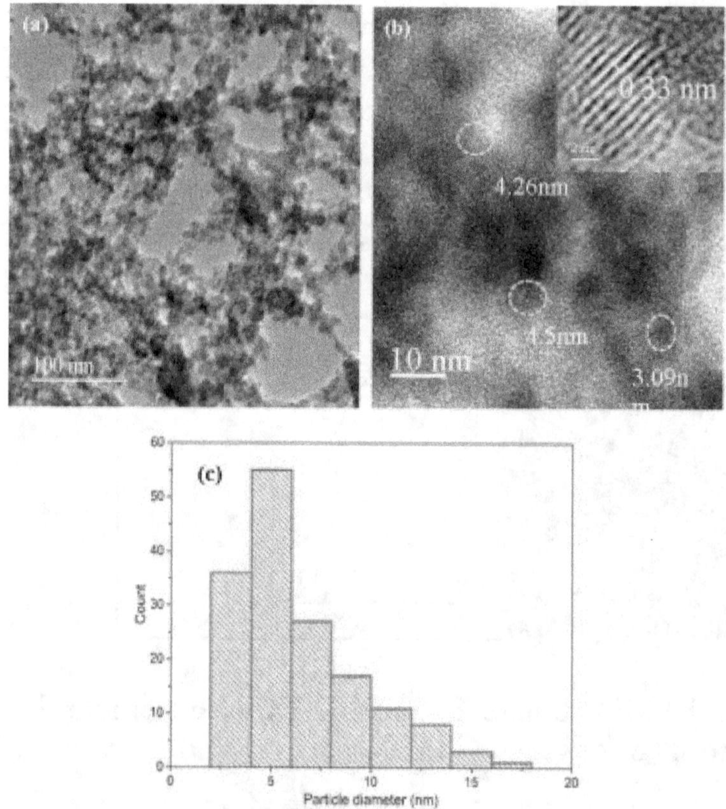

Figure 3.3 TEM (a) HRTEM (b) Histogram (c) for particle size distribution of green CdS QDs. Inset of Fig.b shows single QDs

Optical Studies of Green CdS Quantum Dots

The UV-Vis optical absorption spectrum and corresponding Tauc plot of green CdS QDs are shown in Figure 3.4 (a and b). The optical absorption spectrum shows a strong absorption peak centered at 433 nm due to band edge emission of CdS QDs. As can be seen from the Figure 3.4 (b), the band gap of green CdS QDs is found as 2.85 eV. The Tauc plot implies that green CdS QDs have higher band gap than that of bulk CdS ($\sim$2.4 eV). The widening of band gap were originated from the quantum confinement effect of CdS QDs as the absorption peak get blue shifted in Figure 3.4 (a) compared to bulk CdS ((Ha *et al.* 2014; He *et al.* 2015).

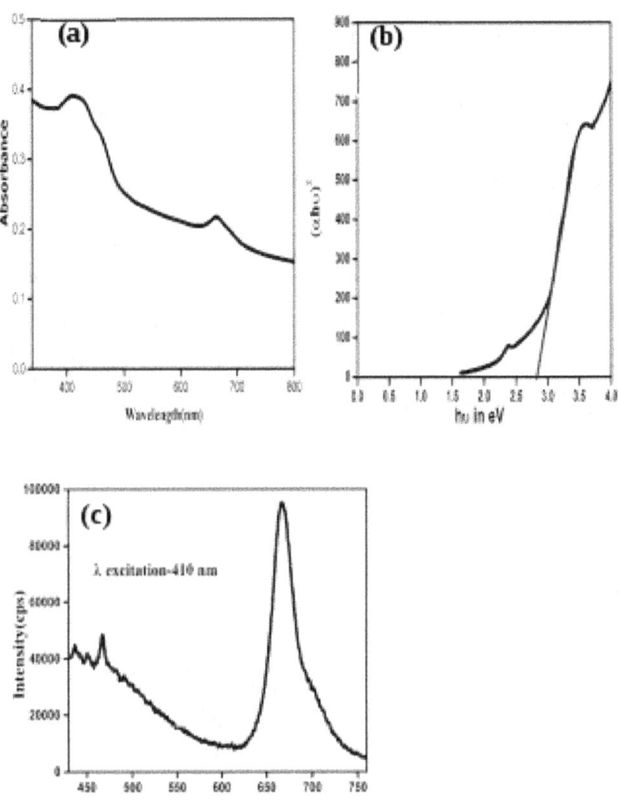

Figure 3.4 UV-Vis absorption spectrum (a) tauc plot (b) fluorescence spectra (c) of green CdS QDs

Figure 3.4 (c) shows the fluorescence spectrum of green CdS QDs. An excitation wavelength of 410 nm was fixed for the fluorescence analysis. As prepared green CdS QDs exhibited strong fluorescence in the range of 625-750 nm, with a maximum intensity at 670 nm. In addition, weak emission was observed at 430 nm which corresponds to band edge emission of CdS QDs. Further two peaks were observed in the range of 450 to 470 nm which might be due to emission from the shallow level surface defects in the CdS QDs (Shivaji et al. 2018).

FTIR of Green CdS Quantum Dots

The FTIR spectra is recorded for A. indica plant extract Figure 3.5

(a) and green CdS Figure 3.5 (b). The IR spectrum of A. indica extract showed double absorption bands centered at 3420 cm-1 and 3183 cm-1 due to stretching vibrations of primary and secondary amines of plant extract (Janakiraman *et al.* 2015). The peaks at 2948 cm-1 and 2850 cm-1 represent the aliphatic C-H stretching of methanol solvent. The peak at 1653 cm-1 was due to stretching of C=O from ketone groups of A. indica and the peak at 1001 cm-1 was owing to the presence of aliphatic amines.

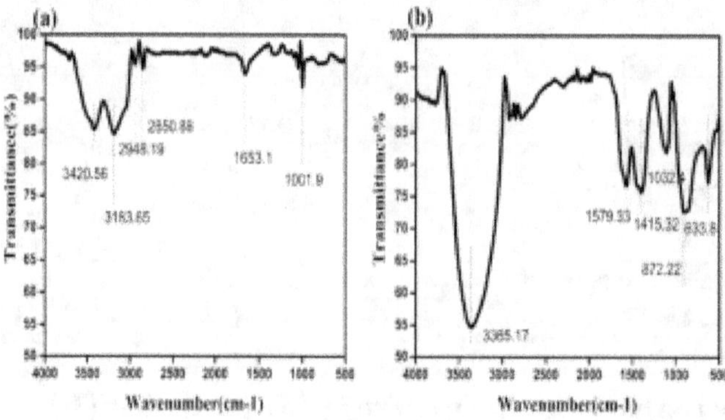

Figure 3.5 FTIR of (a) Azadirachta Indica extract and (b) green CdS QDs

FTIR peaks of green CdS were laid out in Figure 3.5 (b), includes the band at 633 cm-1 which was the characteristic peak of Cd-S stretching and confirming the formation of CdS QDs. Presence of few organic functional groups were also observed in FTIR spectrum of green CdS at 1415 cm-1 and 1032 cm-1 due to C-C and C-O stretching, respectively. The FTIR spectrum of green CdS shows a broad peak at 3365 cm-1 due to O-H stretching, and a peak at 1579 cm-1 was because of O-H bending.

Formation Mechanism of Green CdS Quantum Dots

Growth mechanism of green CdS QDs were explained by considering the interaction between the organic molecules of plant extracts and CdS surfaces (Padian *et al.* 2011). Plants extracts have the potential to hyper accumulate and biologically reduce ions. A. indica extract consists of polyphenols, proteins, caffeine, amino acids (Padian *et al.* 2011). Protein binders within the plant extract plays major role in controlling size of CdS shape-control modifier and stabilizing agent (Djibril *et al.* 2015).

At the initial stage, during the addition of cationic source (cadmium chloride) to plant extract, Cd_{2+} ions bind with plant mediated protein due to in- situ metallic stress from bioactive molecules present in plant extract. On further addition of Na_2S, it facilitates formation of S_{2-} ions and binds to plant mediated Cd_{2+} ions and form CdS QDs. Bioactive components like amines, carboxylic acids, ketones, hydroxyl groups as identified from FTIR spectrum of plant extract in Figure 3.5 (a) plays dominant role in reducing cadmium and then stabilizing them by interacting with sulphur and forming CdS with controlled sizes and shapes (Djibril *et al.* 2015).

When organic components or alien ions are attached to growth surface strongly, the available growth sites are occupied, thereby growth process terminates. Presence of small amount of organic residuals with green CdS QDs can be seen from FTIR Figure 3.5 (b) which can act as an organic capping and reducing its toxicity (Pan *et al.* 2014).

Toxicity Study

Hemolysis assay was performed to evaluate the hemocompatibility of the green CdS QDs and shown in Figure 3.6. Procedure for hemolytic study can be explained as follows: Healthy human blood sample 2 mL freshly harvested in a vacutainer tubes containing anticoagulant. The serum was separated and the RBCs were suspended in 5 mL phosphate buffered saline (PBS, pH 7.4) and washed thoroughly thrice with PBS. Then 0.2 mL of RBCs were pipetted into 1.5 mL microcentrifuge tube containing different concentrations of CdS QDs (50,100, 250, 500, 1 µg/mL). The samples were incubated at room temperature (24 °C) for 2 h. The samples were centrifuged at 10000 rpm for 2 min, the supernatant was finally taken out for recording optical absorbance analysis at 541 nm wavelength. Note that the RBCs were incubated with water and PBS for positive and negative controls, respectively. The percentage of hemolysis (%) is estimated using the relation

$$\% \ \text{Hemolysis} = \frac{\text{Sample absorbance-negative control}}{\text{Positive control-negative control}} \times 100$$

Figure 3.6 (a, b) shows the effect of green CdS QDs on RBC integrity. Figure 3.6 (a) shows the percentage hemolysis of various concentration of green CdS QDS represented as bar diagram. Figure 3.6 (b) shows the photograph of RBC after treatment with various concentration of green CdS. Positive control shows control shows intact RBC and negative control on complete lysis of RBC. Green CdS QDs shows the toxicity value of

0.026 % which seems to be very low (Dobrovolskaia *et al.* 2008).

Further toxicity was analyzed by studying cytotoxic effect of green CdS QDs in comparison with CdS QDs (Shivaji *et al.* 2018).

Toxicity was studied on human lung epithelial cells (A549) and the results were shown in Figure 3.7 (a-d). Figure 3.7 (a-c) gives phase contrast image of cells captured

% Hemolysis of Green CdS using Lieca DME8 microscope at 40X magnification. The cell morphology at various treatments of QDs was also provided in Figure 3.7 (a-c). The cell morphology was significantly altered in CdS QDs treated cells compared to green CdS QDs.

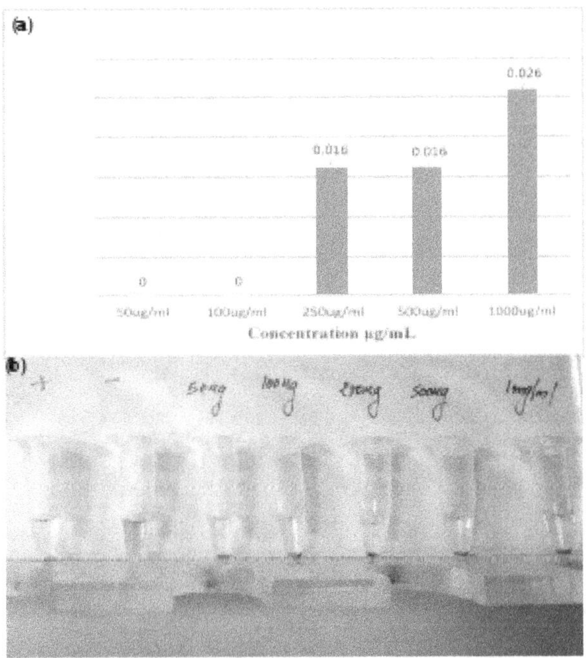

Figure 3.6 Hemolytic analysis to study effect of green CdS on RBC integrity (a) Bar diagram representing percentage hemolysis at various concentration of green CdS (b) Photograph of RBC after treatment with various concentration of green CdS

The results are shown in Figure 3.7 (d). The results indicate that green CdS QDs does not induce any cytotoxic effect in cell at concentration as high at 100 μg when compared to CdS QDs. The cell viability was only 64 % in cell treated with CdS QDs at 50 μg when compared to green QDs which showed cell viability of 97 %. The cell viability was only 4 % in CdS QDs at

100 µg concentration when compared to green CdS QDs which showed a viability of 80 % at the same concentration. Hence the results indicated that green CdS QDs does not induce any toxic effect on human cell lines and hence can be considered as nontoxic in nature.

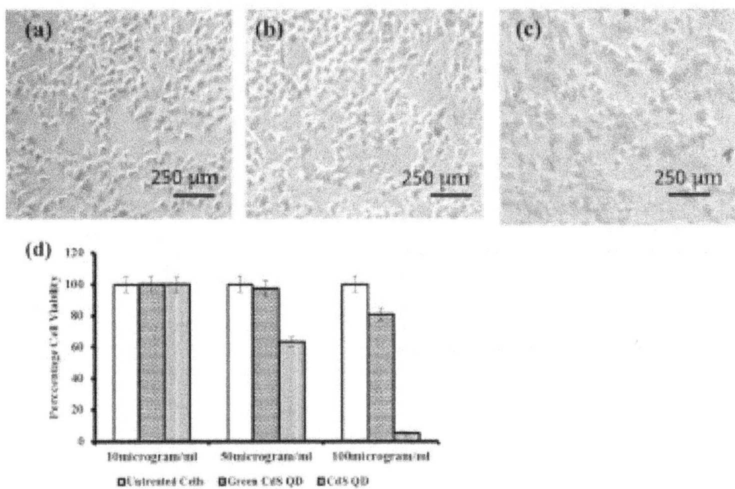

Figure 3.7 Phase contrast images of cells (A54) (a) untreated cells (b) cells treated with green CdS QDs (c) cells treated with CdS QDs (d) percentage cell viability

Photoanode Characterization

Figure 3.8 (i) and (ii) shows the XRD patterns of green CdS loaded anatase and rutile TiO_2 photoanodes, respectively. Figure 3.8 (iii) and (iv) are the XRD patterns of SILAR deposited CdS on anatase and rutile TiO_2 photoanodes respectively. The typical diffraction peaks with (101), (200) and

(211) peaks corresponding to anatase TiO_2 matches with JCPDS card No.21- 1272 (Figure 3.8 (i) and (iii)). The high intensity peaks

at (101), (400) and (420) of rutile TiO_2 matches with JCPDS card no: 21-1276 (Figure 3.8 (ii) and (iv)). Green CdS and SILAR CdS were well matched with standard JCPDS data

(No.10-454). The peaks observed at 2θ of 26.4,30.2, 43.6, 51.9 and 70.4

corresponds to (111), (200), (220), (311) and (331) planes of the face centered cubic CdS nanoparticles. Further, CdS related diffraction peaks were well matched with JCPDS card no: 10-454 that confirms the effective loading of QDs into the photoanodes with different phases. Additionally, peaks corresponding to FTO substrate at 2θ of 25.9, 56.3,68.1 and 79.2 can be clearly observed in the XRD patterns.

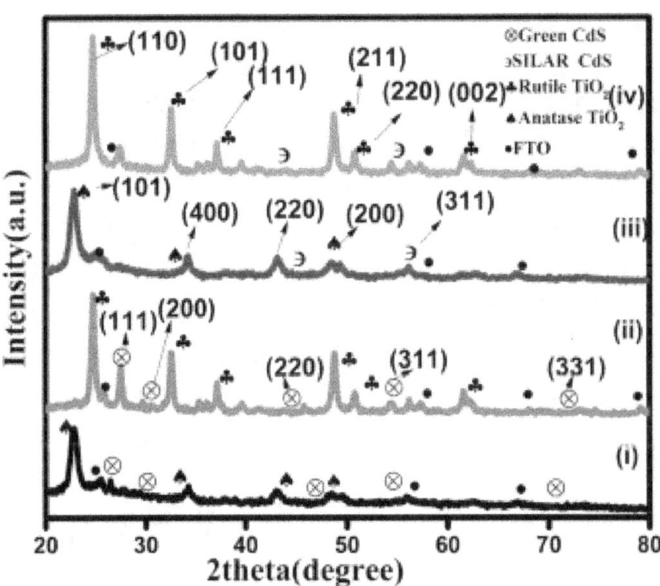

Figure 3.8 XRD patterns of (i) green CdS on anatase TiO₂ coated FTO

(ii) green CdS on rutile TiO₂ coated FTO (iii) SILAR deposited CdS on anatase TiO₂ (iv) SILAR deposited CdS on rutile TiO₂

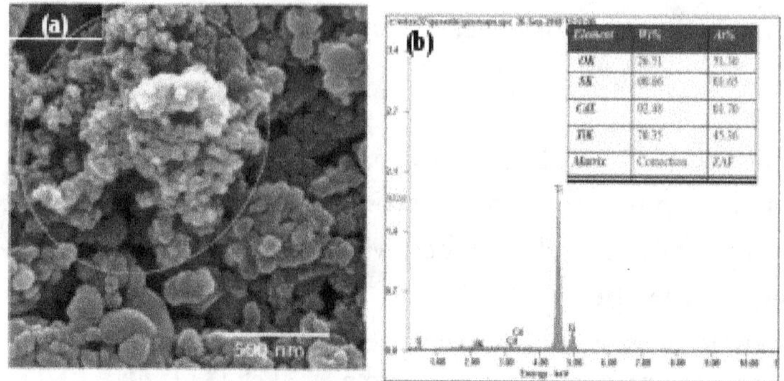

Figure 3.9 SEM images (a) and EDAX spectrum with elemental composition (b) of green CdS QDs loaded TiO$_2$ photoanode

Figure 3.9 (a and b) shows the FESEM image and EDAX spectrum of green CdS QDs loaded TiO$_2$ photoanode, showing effective loading of CdS QDs on the photoanode and the QDs are indicated as yellow colour dotted circle in the image in Figure 3.9 (a). Moreover, the size of the QDs was extremely low compared to TiO$_2$ particles in the photoanodes. The EDAX spectrum in Figure 3.9 (b), shows the presence of Cd and S in addition with Ti and O, which confirms the presence of CdS QDs in the photoanode. The chemical composition of the photoanode was given as inset of Figure 3.9 (b).

UV-Vis absorption spectra of green CdS and SILAR deposited CdS loaded rutile and anatase TiO$_2$ photoanodes are shown in Figure 3.10. The high absorption in visible regime of green CdS loaded photoanode favouring larger absorption window thereby it was expected to have better QDSSC performance (Gao *et al.* 2016). Moreover, green CdS loaded TiO$_2$ photoanode showed relatively high intensity of absorption compared to SILAR deposited CdS. The presence of auxochromes in organic moieties of green CdS increases the intensity of absorption. Moreover, hypsochromic shift can be seen for green CdS loaded photoanode compared to SILAR

CdS owing to confinement effect of green CdS QDs loaded TiO_2 photoanode.

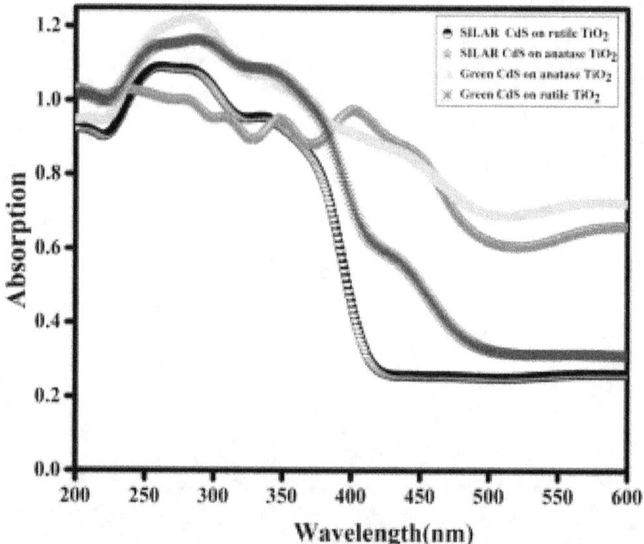

Figure 3.10 UV-Vis absorption spectra of green CdS loaded anatase and rutile TiO₂ photoanodes

J-V and EIS Characteristics of fabricated QDSSCs

The fabricated QDSSCs is tested with standard solar simulator under 1 sun illumination condition. The obtained cell parameters are shown in Table

Figure 3.11(a-c) unfold schematic representations of QDSSC with different configurations such as SILAR CdS sensitized QDSSC, green CdS sensitized QDSSC and green CdS based QDSSC with ZnS passivating layer, respectively.

Figure 3.12 (a and b) shows the J-V curves of green CdS and SILAR CdS based QDSSCs fabricated using rutile and anatase TiO_2 as photoanodes and Pt as CE. As can be seen from Figure 3.12 (a), the QDSSC with anatase TiO_2 photoanode showed relatively better performance compared to rutile TiO_2 photoanodes for both kind of green and SILAR CdS QDs.

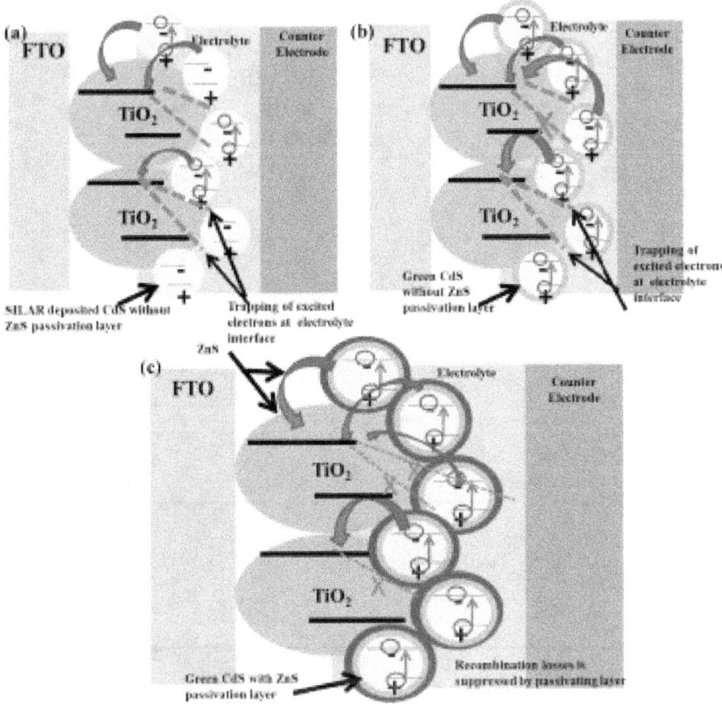

Figure 3.11 Schematic representation of QDSSC with (a) SILAR CdS (b) Green CdS (c) Green CdS with ZnS passivating layer

The relatively high performance of anatase TiO_2 photoanodes based QDSSCs were possibly due to the shift in the Fermi level by 0.1 eV compared to rutile phase resulting in a lower oxygen affinity and a higher level of hydroxyl groups on TiO_2 surfaces which favours carrier transport (Bai *et al.*2014). Anatase possesses an indirect bandgap while rutile has a direct band gap thus minima of the conduction band of anatase is away from the maxima of the valence band, enabling the excited electrons to stabilize at the lower level in the conduction and leading to a longer life (slower charge carrier recombination) and higher mobility than that of rutile with a direct band gap. Also, anatase TiO_2 possesses a wider band gap.

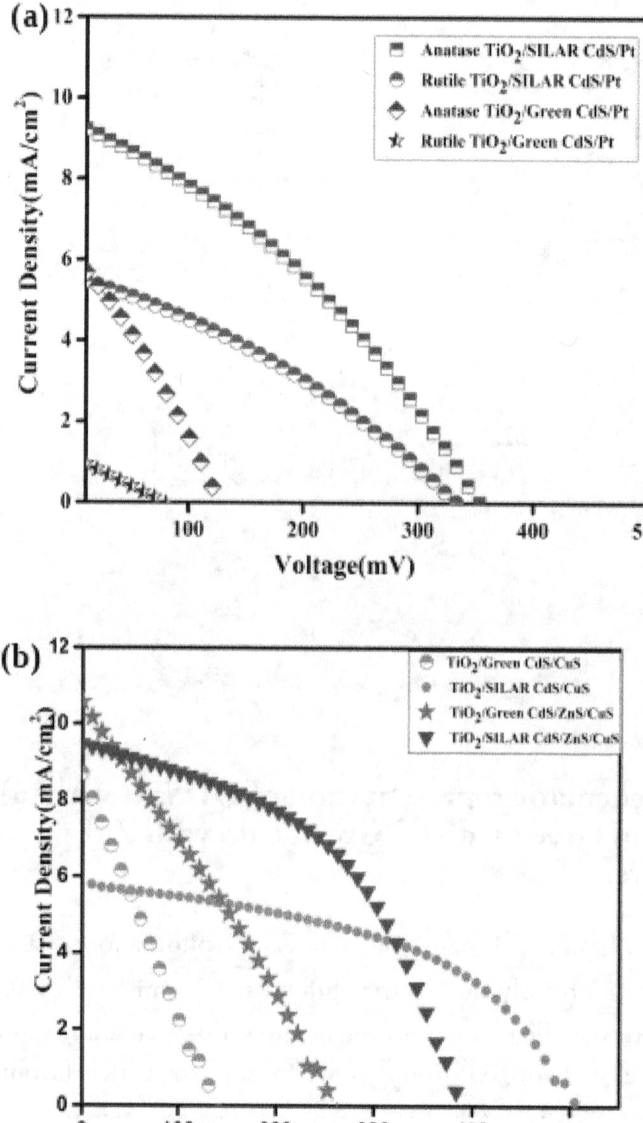

Figure 3.12 J−V characteristics of fabricated QDSSCs with (a) Pt Counter Electrodes (b)CuS CounterElectrodes.

Excitation electron mass of the outer shell electrons of anatase is lower than that of rutile, leading to a higher mobility of electrons in anatase TiO_2 (Bharti *et al.* 2016). Hence improved performance can be observed for anatase phase TiO_2 photoanode with both green CdS and SILAR deposited CdS (Kumar *et al.* 2011; Liu *et al.* 2019). Therefore, the anatase TiO_2 based photoanodes were used for further improving the QDSSC performance by employing the passivating layers and replacing the CE.

Figure 3.12 (b) shows the J-V curves of green and SILAR CdS loaded anatase TiO_2 based QDSSCs with CuS as CE. Moreover, the J-V characteristics of the QDSSCs were analyzed with and without ZnS passivating layers. As reported in literature, passivation layer helps exciton generation in the core to remove defects, such as unsaturated surface atoms on the surface, and to reduce alternative decay pathways and reducing the charge recombination rate at the interfaces (Kim *et al.* 2015b). Hence an improved performance can be seen with ZnS layer coating. As can be seen from Figure 3.12 (b), the performance of QDSSC with CuS as CE were relatively high compared to the QDSSC with Pt CE Figure 3.12 (a). The sulphur from the polysulphide electrolyte may chemisorbs on to the Pt electrode surface, which resulted higher R_{ct} at the interface of CE/electrolyte and lowers the performance. Whereas, the ZnS was more compatible with polysulphide electrolyte thereby it shows relatively high performance (Figure 3.12 (b)). From the J-V curves Figure 312 (b) it can be inferred that the J_{sc} of green CdS based QDSSC was relatively higher (10.61 mA/cm_2) compared to SILAR CdS based cell (9.2 mA/cm_2) because of higher visible photon absorption as can be seen from UV-Vis absorption spectra Figure 3.12 (b).

The presence of organic moieties with green CdS causes suppression of surface defects and improving flocking of charge carriers. The broadened light absorption from 250 to 520 nm with high intensity favors light harvesting efficiency and presence of retral organic molecules effectively encapsulate the QDs and thereby improved electron hole injection that eventually improved J_{sc} of green CdS based QDSSC. An overall abate in open circuit voltage (V_{oc}) results in lowering of photo conversion efficiency of green CdS based QDSSCs. The decay in V_{oc} was mainly attributed due to recombination losses that deliberately downturn excitonic life time of charge carriers. A significant improvement of 65 % in V_{oc} was observed for green CdS QDSSC after passivating with ZnS. From the results, it was obvious that the ZnS layer preventing retension of electrolyte towards sensitizer and to TiO_2 photoanode layer that can expedite the recombination of the charge carriers. This phenomenon was schematically shown in Figure 3.11 (c). Although more photo-generated carriers were generated in the green CdS, the likelihood of electron-hole recombination was still high. Consequently, the photo conversion efficiency of the green CdS based QDSSCs were lower (0.77 %) compared to SILAR deposited CdS (1.93 %). Hence much better control of recombination process has to be done for achieving higher conversion efficiency in green CdS QD sensitized solar cells.

Table 3.1 Photovoltaic characteristics of QDSSCs with different combinations

Cell Configurations	V_{oc} (mV)	J_{sc} (mA/ cm$_2$)	FF	Efficiency (%)
Rutile TiO$_2$/green CdS/Pt	28.7	4.73	0.23	0.02
Anatase TiO$_2$/green CdS/Pt	115.1	5.80	0.28	0.19
Rutile TiO$_2$/SILAR CdS /Pt	343.5	5.90	0.32	0.66
Anatase TiO$_2$/SILAR CdS /Pt	352.8	9.36	0.33	1.12
Anatase TiO$_2$/green CdS/CuS	169.0	10.61	0.24	0.44
Anatase TiO$_2$/SILAR CdS/ CuS	495.9	9.21	0.47	1.36
Anatase TiO$_2$/green CdS/ZnS / CuS	258.7	10.56	0.28	0.77
AnataseTiO$_2$/SILAR dS/ZnS/ CuS	419.3	9.66	0.47	1.93
CdS (CBD) on TiO$_2$	355.0	2.57	0.18	0.17
CdS (SAM) on std TiO$_2$(p25)	529.0	0.17	0.59	0.05

EIS analysis were used to evaluate the internal resistance and charge transfer kinetics of QDSSCs. Figure 3.13 (a and b) shows the Nyquist plots of green CdS deposited CdS based QDSSCs with Pt and CuS as CEs and Figure 3.13 (c) shows the SILAR deposited CdS based QDSSCs with CuS as CEs. The plots were analysed using EC lab software and the obtained parameters from the EIS analysis were summarized in Table 3.2. From EIS data, it was clear that the R_s values were high for QDSSCs with Pt CE compared to that with CuS CEs (Liu *et al.* 2019; Kumar *et al.* 2019b; Gonzalez-Pedro *et al.* 2010; Yu *et al.* 2010).

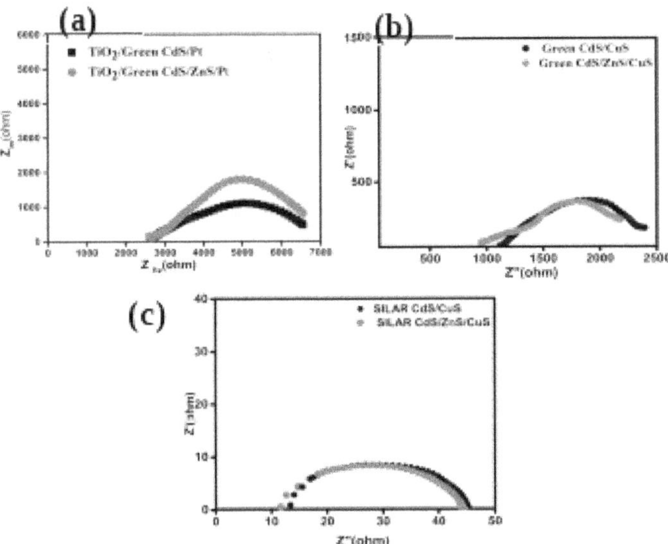

Figure 3.13 Electrochemical Immpedence spectra of QDSSC fabricated with and without ZnS passivation layer using green CdS with

(a) Pt CE (b) using CdS with CuS CE (c) using SILAR CdS with CuS CE

The lower R_{ct} value shows higher electrocatalytic activity of CuS based QDSSCs. Relatively higher R_{ct} values of Pt based QDSSC shows the incompatibility of the cell with sulphur based electrolyte which was used in the QDSSC. Therefore, the QDSSCs with CuS as CE showed relatively higher photocurrent compared to cell with Pt CE. Moreover, the QDSSC with ZnS passivating layer showed relatively high R_{ct} compared to the cell with no passivating layer. The EIS data revealed that the QDSSC with CuS CE and ZnS passivating layer as an optimum configuration with low R_s and R_{ct} for high photovoltaic performance.

Table 3.2 EIS characteristics of QDSSC fabricated with different combination

Cell Configurations	Rs(ohm)	Rct(ohm)
TiO$_2$/Green CdS/Pt	2657	231.41
TiO$_2$/Green CdS/ZnS/Pt	2589	331.85
TiO$_2$/Green CdS/CuS	908.2	207.9
TiO$_2$/Green/ZnS/CuS	495.4	217.5
TiO$_2$/SILAR CdS/CuS	13.87	11.74
TiO$_2$/SILARCdS/ZnS/CuS	13.8	15.7

CONCLUSION

CdS QDs were synthesized using A. indica as a stabilizing agent and its potential ability to enhance the performance of QDSSC were investigated in comparison with conventional SILAR CdS. Green CdS QDs based QDSSCs with ZnS passivating layer and CuS CE showed relatively higher J_{sc} (10.61 mA. cm$_{-2}$) compared to other combinations of cells. The UV-Vis absorption spectra revealed the improved optical absorption especially in the visible region for green CdS QDs compared to SILAR CdS QDs based QDSSCs. Surface modification by ZnS passivating layer on green CdS suppress recombination thereby improving the V_{oc} and photo conversion efficiency of green CdS based QDSSC. Further, the experimental results demonstrated that green CdS QDs were favorable as a sensitizer for QDSSCs, as it exhibits the efficiency of 0.76

% with high J_{sc} of 10.61 mA.cm$_{-2}$. Key research challenges like toxicity of chalcogenides were controlled by effective green synthesis approach resulting QDs with presence of organic moieties without necessarily modifying its electronic property.

CHAPTER 4

STUDIES ON InSb QUANTUM DOTS AS CO-SENSITIZER FOR CdS BASED QUANTUM DOT SENSITIZED SOLAR CELLS

INTRODUCTION

The low efficiency of QDSSCs are partially associated with the performance of semiconductor based sensitizer materials which are mostly active only in the visible regime of solar light spectrum and hence light absorption get confined at wavelength less than 700 nm (Maurice *et al.* 2013; Yaremaa *et al.* 2013). Co-sensitization of QD materials with different bandgap is a promising approach for enhancing the performance of the QDSSC by covering wide range of solar spectrum beyond the visible regime (Hatami *et al.* 2006; Lin *et al.* 2017). In this aspects, Q. Tian *et al.* (2017). reported co-sensitization of N719 dye with Ag_2Se QDs showed an efficiency of 3.23% . Lin *et al* (2017*).* reported co-sensitization of CdS and CdSe QDs with an improved efficiency of 4.2%. D. Liu *et al.* (2019) reported 3.95%, efficiency for the co- sensitization of CdS and CdSe QDs on a comparative study using Pt and CuS CE.

Hwang *et al* (2013). reported the significant improvement of Jsc (27.6 mA / cm_2) of QDSSC by co-sensitization of Ag_2S with CdS QDs . From the literature, it is obvious that the co-sensitization is a promising approach compared to single sensitizer (CdS or PbS or CdSe) for improving the performance of QDSSCs (Balis *et al.* 2013). Tung *et al.* reported the controlled recombination achieved by co-sensitized structure (Tung *et al.* 2018). Most of the literatures highlight on visible active materials as co-sensitizer for sensitized solar cells with an objective to improve the eiffciency (Wang *et al.* 2019). Whereas, the visible (44 %) and infrared (53 %) radiations dominate the solar spectrum which covers from 280 to 2700 nm (Muthalif *et al.* 2017, Tasco *et al.* 2007; Vogel *et al.* 2010). Therefore, effective utilization of IR region by means of IR active QDs could enhance the efficiency of QDSSC. However, co-sensitization of IR active QD material with visible active QD materials are rarely investigated for solar cell applications.

Indium antimonide (InSb) is one of the narrow band gap (0.17 eV) semiconductors with large excitonic bohr radius of 68.5 nm. Therefore, the properties of InSb QDs can be tuned by changing its size in the confinement regime (Tian *et al.* 2017; You *et al.* 2019; Sukkabot *et al.* 2014). Moreover, InSb has been reported to be an efficient IR active material as it possesses narrow band gap (Pan *et al.* 2012; Wang *et al.* 2015b). Therefore, co- sensitization of InSb with visible active QDs could enhance the quantum efficiency of QDSSC. However, co-sensitization of InSb QDs with other semiconducting QDs are not reported in the literature. Henceforth InSb can be studied as viable co-sensitizer material for QDSSC applications

Hitherto InSb QDs and its applications are scarcely ever considered as they are highly reactive towards air and moisture due to highly reactive nature of group III–V elements (Li *et al.* 2011; Atoyan *et al.* 2011; Algarni *et al.* 2016; Zuhair *et al.* 2009). Most of the works reported on in-situ growth of InSb nanoparticles and deposition of InSb as thin films (Hnida *et al.* 2016; Pandya *et al.* 2015; Shafa *et al.* 2016; Singh *et al.* 2017). Only few works have been reported on the synthesis of InSb QDs (Vogel *et al.* 2010; Li *et al.* 2001). Therefore, in this chapter, the conditions for the formation of InSb QDs were initially optimized and examined its suitability as a photosensitizer for QDSSC with and without co-sensitization of CdS. The co-sensitized $TiO_2/CdS/InSb$

QDSSC exhibited high efficiency of 4.94 %, which is 40 % higher than that of TiO_2/CdS (3.52 %) based QDSSC.

EXPERIMENTAL

Synthesis of InSb Quantum Dots

High purity indium particles (99.99%) (Merck), antimony trichloride ($SbCl_3$) (TCI Chemicals), xylene (Alfa Aesar) and ethanol (Honeywell) were procured and used as source materials without further purification.

InSb QDs were synthesised by solvothermal self-reduction method using high purity In metals and $SbCl_3$ as precursors. In and $SbCl_3$ were taken in a molar ratio of 1: 2 and dissolved in 100 mL of xylene in polypropylene lined (PPL) vessel filled with xylene up to 70 % of the total volume. The vessel was sealed into a stainless-steel tank and maintained at 280 °C for 15 h. After completion of solvothermal reaction, autoclave was allowed to cool down to room temperature and resulted product was filtered off, washed with double distilled water and absolute ethanol. The resulted blackish brown product was isolated by centrifugation at 8000 rpm and transferred to petridish and vacuum dried at 80 °C for 3 h. The obtained InSb nanostructures were characterized for structural, morphological and optical analysis.

Characterization of InSb Quantum Dots

The structural properties of the InSb samples obtained by solvothermal synthesis were examined by X-ray diffraction (XRD) analysis using X-ray diffractometer (Bruker D8 advance Diffractometer). The UV-Vis- NIR absorption spectrum of InSb was recorded using PerkinElmer Optima 5300 DV UV-Vis-NIR spectrophotometer. XPS analysis of prepared samples was carried out by Shimadzu ESCA 3400 X-ray photoelectron spectrophotometer.

Fabrication of QDSSC

a. Preparation of photoanode.

Titaniumtetraisopropoxide, chloroplatinic acid (Sigma Aldrich), nitric acid, acetic acid, acetone, copper acetate, cadmium chloride (Alfa Aesar), ethanol (Honeywell), FTO (7 Ω resistance, Sigma Aldrich), ethyl cellulose, α- terpinol, Na_2S (TCI Chemicals) were procured and used as source materials without further purification.

The TiO_2 nanoparticles were prepared from TTIP, acetic acid via a hydrothermal method at temperature of 240 °C for 12 h (Ito *et al.* 2007). After hydrothermal reaction 0.1 M of HNO_3 was added to neutralize the pH and resulted product were washed three times using double distilled water. The resulted white product was isolated by centrifugation at 6000 rpm and transferred to petridish. After that the sample was dried at 100 °C and white TiO_2 powder was obtained. FTO substrates were well cleaned with detergent and washed with water and ethanol. The photoanodes were prepared by coating TiO_2 paste of respective samples on FTO substrate using doctor blade technique. 2 g of prepared TiO_2 powder was mixed with 0.5 mL of acetic acid and mechanically grinded for 10 min for preparing TiO_2 paste.

After that 0.5 mL of double distilled water was added and continued the grinding. Further, 2 mL of ethanol and 2 mL of α-terpineol were added and continually grinded for 45 min. Further 0.5 g of ethyl cellulose and 6 mL of ethanol were added, while grinding continued for 60 min (Chang *et al.* 2014). The paste was coated over conducting side of FTO substrate for area of 0.16 cm_2. The prepared photoanode was dried at 100 °C for 30 min and calcined at 450 °C for 15 min and slowly cooled down to room temperature.

b. InSb QD loading onto TiO_2 photoanode

InSb QDs were sonicated in ethanol for 30 min and loaded on TiO_2 photoanode by direct absorption method. QD loaded photoanode was dried by purging N_2 gas. To prepare the co-sensitized photoanode CdS QDs were loaded by SILAR deposition of 0.04 M concentration of cadmium and sulphur source for 8 cycles. (Mikhailov *et al.* 2017; Subash *et al.* 2014) Over which InSb QDs were loaded by direct absorption, followed by drying with N_2 gas purging for 5 min (Tamang *et al.* 2015; Vaemsunthorn *et al.* 2017; Wang *et al.* 2017).

c. Preparation of Counter electrode

Initially FTO substrates were well cleaned and immersed in isopropanol and sonicated for 30 min. After that, washed with double distilled water and ethanol for two times. For a comparative analysis Pt and CuS coated counter electrodes (CE) were prepared separately. In a typical process of Pt coated CE, a drop of H_2PtCl_6 solution (20 mM of chloroplatinic acid in isopropyl alcohol) was dropped onto FTO substrate, followed by heating at 450 °C for 30 min (Li *et al.* 2017; He *et al.* 2015). SILAR method was used to prepare CuS CE on the FTO substrate. The CuS CE was prepared by applying 4 cycles of SILAR deposition using cationic and anionic aqueous solutions of

0.5 M Cu $(NO_3)_2$ and 0.5 M Na_2S. Finally, the CuS CE was dried at 120 °C for 10 min (Kalanur *et al.* 2013; Lee *et al.* 2010).

d. Preparation of electrolyte and fabrication of QDSSC.

Polysulphide electrolyte was prepared by adding 1.5 M KCl, 2 M S and 1.5 M Na_2S in 10 mL of double distilled water. The photoanode and CE were sandwiched together and the space between the electrodes was filled with prepared electrolyte (Wang *et al.* 2017). J–V characteristics of the prepared QDSSC were studied by solar simulator (1 Sun Oriel Class AAA). EIS of fabricated QDSSC were recorded by CHI electrochemical workstation (CHI 1106C) at frequency ranging from 10 MHz to 100 kHz and impedance spectra were analysed with EC lab software.

RESULT AND DISCUSSION

Structural and Morphological Analysis

The X-ray diffraction (XRD) pattern of InSb nanoparticles is shown in Figure 4.1. The diffraction pattern was well matched with cubic zinc blende structure of InSb. (Hnida *et al.* 2016; Tamang *et al.* 2015) In addition, a few peaks with low intensity related to Sb was observed due to excess amount of Sb taken during the synthesis (Yaemsunthorn *et al.* 2017). The obtained cell parameter (a = 6.4781) was in good agreement with the standard JCPDS data

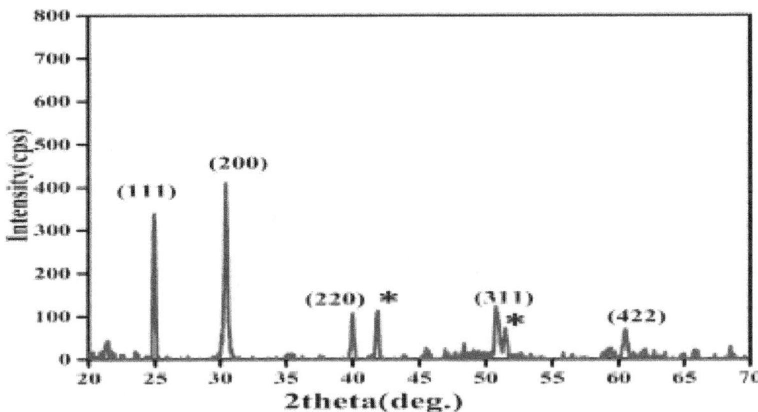

card no. 6-208. From the XRD pattern shown in Figure 4.1, the size of the crystallites of InSb particles was calculated to be 37.16 nm using Scherrer equation.

Figure 4.1 XRD pattern of InSb

Morphology of InSb nanoparticles was studied by FESEM and shown in Figure 4.2 (a). Spherical morphology of the InSb nanoparticles can be seen from FESEM images and Figure 4.2 (b) shows EDAX spectrum of the sample. The strong peaks corresponding to In and Sb were clearly observed. The composition of the prepared InSb was measured by EDAX spectrum shown in Figure 4.2(b) with composition ratio of elements were shown as inset of Figure 4.2 (b).

(a) **(b)**

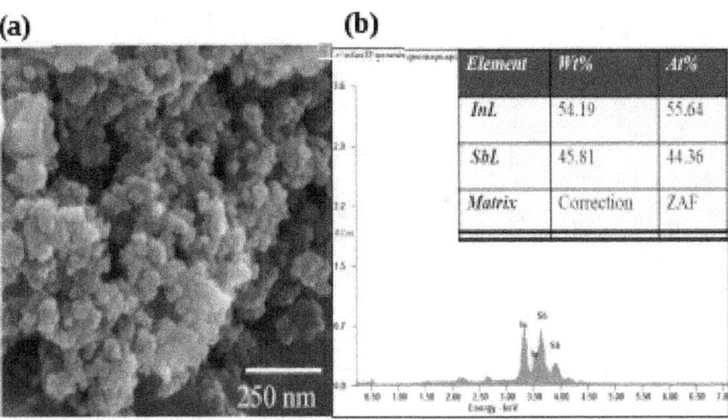

Element	Wt%	At%
InL	54.19	55.64
SbL	45.81	44.36
Matrix	Correction	ZAF

Figure 4.2 SEM image (a) and EDAX spectrum with elemental composition (b) of InSb QDs

Typical HRTEM images of the InSb samples are shown in Figure 4.3 (a) and (b). The size and morphology of InSb particles can be clearly identified using TEM images. The nanocrystalline InSb had plate-like and spherical morphology as shown in Figure 4.3 (a) and (b). The average size of InSb nanocrystals was found to be less than 25 nm which confirms the formation of InSb QDs as the size is less than its excitonic Bohr radius (68.5 nm). Inset of Figure 4.3 (a) and (b) gives the histogram for the particles size distribution of InSb and can be confirmed to be less than 25 nm. The HRTEM image in Figure 4.3 (c) of InSb QDs shows the clear lattice fringes which confirm its crystalline nature. The interplanar spacing of InSb QDs were measured at two different places as 0.38 nm and 0.32 nm corresponds to (111) and (200) lattice planes, respectively which were observed as high intensity peaks in the XRD pattern in Figure 4.1.

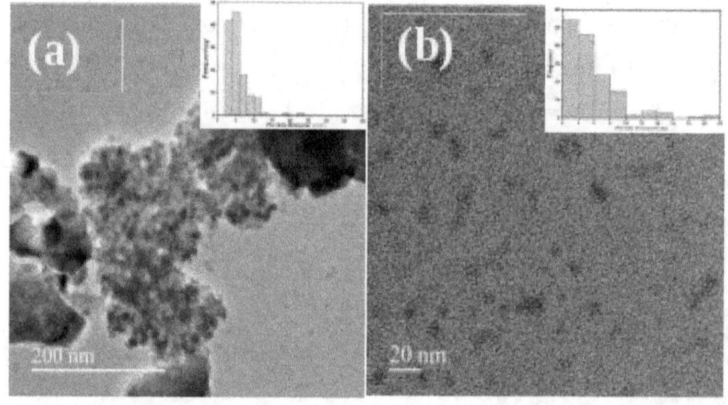

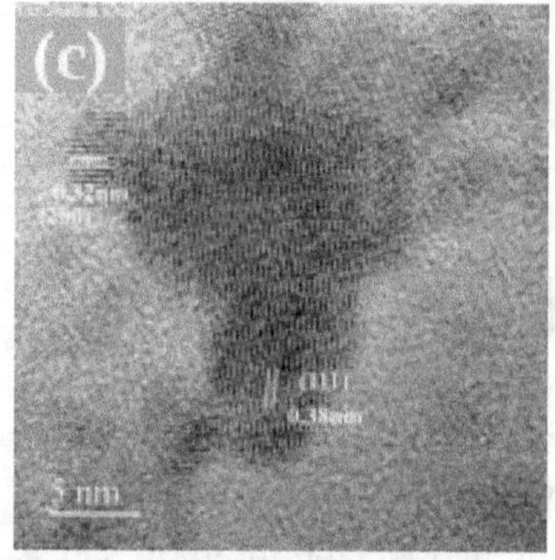

Figure 4.3 TEM images (a and b) of InSb QDs, inset (a and b) are histogram for particle size distribution, HRTEM images (c) of InSb QDs

XPS Analysis of InSb Quantum Dots

The XPS spectra of InSb QDs are shown in Figure 4.4 (a) and (b). The core level spectrum of In (Figure 4.4 (a)) shows the doublet peaks of In $3d_{5/2}$ and In $3d_{3/2}$ at 445.57 and 453.22 eV, respectively. The doublet peaks were separated with the binding energy of 7 eV which arises from spin orbit splitting. The core level spectrum of antimony (Figure 4.4 (b)) shows two peaks at 532.08, and 540.8 eV corresponds to Sb $3d_{5/2}$ and Sb $3d_{3/2}$, respectively. XPS peaks further confirms the formation of InSb. Semiquantitative analysis of the peaks gave the atomic ratio of In to Sb as 52.3: 47.7 (Li *et al.* 2001).

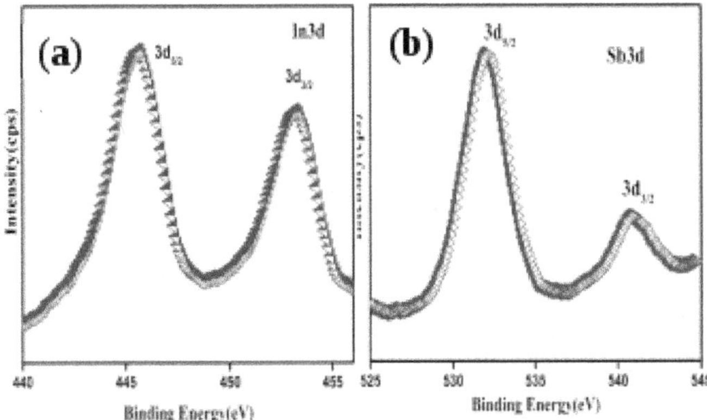

Figure 4.4 XPS Spectra (a) In 3d (b) Sb 3d of InSb Quantum Dots

Optical Studies of InSb Quantum Dots

Figure 4.5 shows the optical absorption spectrum of InSb QDs which reveals a wide range of absorption in the near IR region. From the UV-Vis-NIR spectrum, absorption maximum can be seen at 870 nm and slightly decreasing towards high wavelength. On plotting Tauc plot the estimated bandgap energy is found to be 1.01 eV (inset of Figure 4.5). The band gap for InSb QDs seems to be widened from 0.17 eV to 1.01 eV because of the quantum confinement effect (Subash *et. al* 2014). Optical absorption of InSb QDs shift towards near- infrared thereby can be confirmed to be suitable for improving wide range of light absorption for QDSSC (Katami *et al.* 2006; Tasco *et al.* 2007).

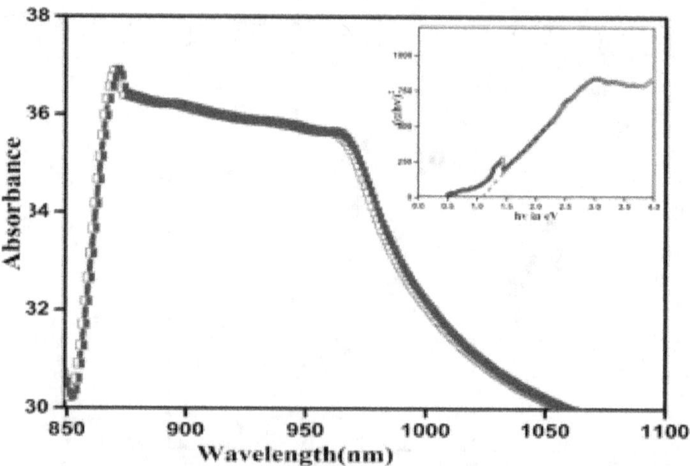

Figure 4.5 UV-Vis-NIR absorption spectrum (inset -Tauc plot) of InSb Quantum Dots

Photoanode Characterization

a. **Structural and morphological analysis of TiO_2/CdS/InSb Photoanode**

Figure 4.6 (a-c) show the cross sectional FESEM images along with EDAX spectrum and mapping for InSb and CdS QDs loaded TiO_2 photoanode. From the FESEM images, it can be clearly observed that the QDs were effectively loaded into the photoanode. Moreover, the size of CdS QDs were extremely low compared to InSb QDs loaded on TiO_2 photoanode. From cross sectional morphological analysis of InSb/CdS/TiO_2 photoanode (Figure 4.6 (a)), stacking of each layer was analysed. From the EDAX spectrum (Figure 4.6 (c)) the presence of In, Sb along with Cd, S, Ti and O, have been confirmed which further illustrates the effective loading of InSb and CdS QDs in photoanodes.

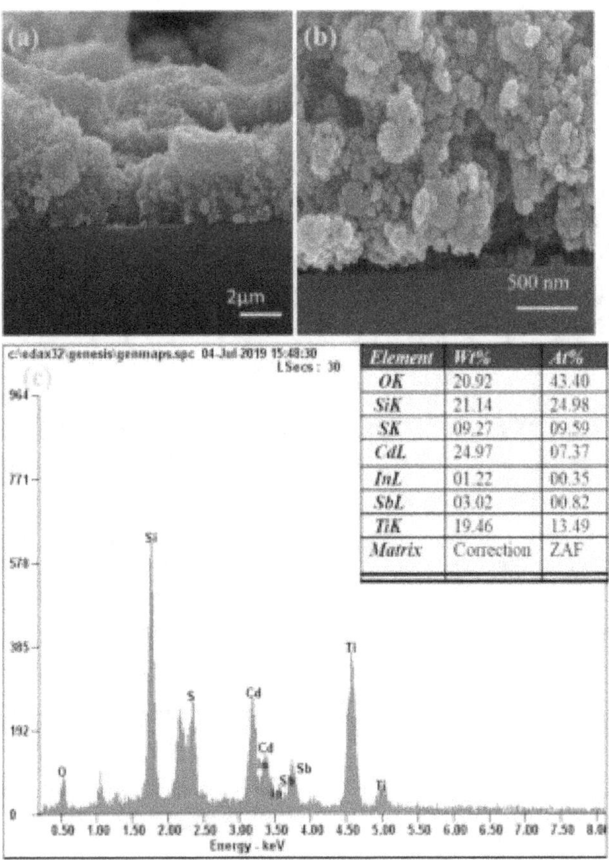

Element	Wt%	At%
OK	20.92	43.40
SiK	21.14	24.98
SK	09.27	09.59
CdL	24.97	07.37
InL	01.22	00.35
SbL	03.02	00.82
TiK	19.46	13.49
Matrix	Correction	ZAF

Figure 4.6 Cross sectional FESEM images (a and b) and EDAX with elemental composition (c) of co-sensitized InSb/CdS loaded TiO$_2$ photoanode

Cross sectional EDAX mapping analysis of InSb/CdS loaded TiO$_2$ photoanode (Figure 4.7) clearly shows the presence of all elements and the composition ratio of InSb were relatively low.

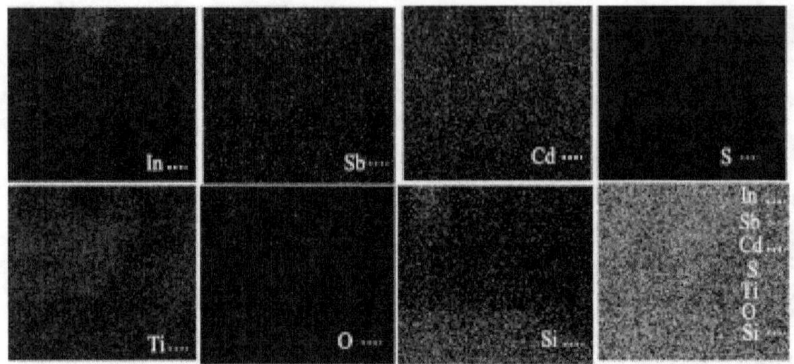

Figure 4.7 EDAX with elemental mapping of cross-sectional surface of co-sensitized InSb/CdS loaded TiO₂ photoanode

Figure 4.8 shows the top view of InSb and CdS QDs loaded TiO₂ photoanode. From the surface analysis of photoanode, higher concentration for InSb QDs is observed (Figure 4.8 (a)). The composition ratio of elements were shown as inset of Figure 4.8 (b). The chemical composition confirms the presence of In, Sb, Cd, S, Ti & O on photoanode.

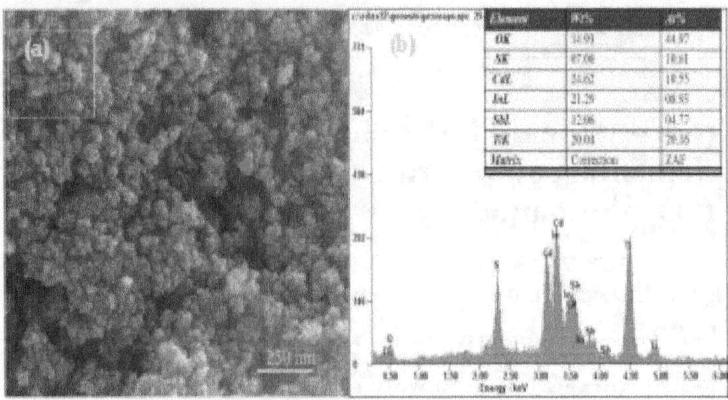

Figure 4.8 FESEM image(a) and EDAX spectrum with elemental composition (b) of surface of co-sensitized InSb/CdS loaded on TiO₂ photoanode

a. Optical study of TiO₂/CdS/InSb Photoanode

Optical properties of photoanode are the significant characteristics that determine the overall performance of the device. Hence a photoanode should have an ability to capture light in wide range of wavelength. UV-Vis- NIR diffuse reflectance spectra of the bare TiO_2, $TiO_2/InSb$, TiO_2/CdS, $TiO_2/CdS/InSb$ are shown in Figure 4.9 (a). Bare TiO_2 photoanode shows absorption in UV region, while on sensitizing it with InSb a hypochromic shift can be seen along with a small hump at 840 nm prominently due to presence of InSb QDs on TiO_2 photoanode. CdS loaded TiO_2 photoanode showed shift in absorption greatly towards visible region along with decrease in intensity of peaks at the UV region due to decrease in surface trap sites of TiO_2 photoanode. On overall analysis of spectrum, the introduction of InSb layer into TiO_2/CdS photoanode broadened the light absorption range especially towards near IR region. Thus, wide range of solar energy harvesting can be achieved by using NIR active InSb QDs along with visible active CdS QDs as sensitizers for QDSSC. This strategy has great potential in improving overall efficiency as it improves overall light absorption with minimal spectral overlap.

Figure 4.9 (b) shows PL spectra of bare TiO_2, $TiO_2/$ InSb, $TiO_2/$ CdS, and $TiO_2/CdS/InSb$ samples. For bare TiO_2 photoanode, two emission peaks were observed at 475 and 540 nm, respectively. The emission peaks in the PL spectra at 475 and 540 nm correspond to band edge and defect level emission of TiO_2. It can be inferred from the spectra that the defect level peaks at TiO_2 were suppressed by QDs loading over the TiO_2 photoanode. The emission peak for CdS QDs was observed at 525 nm to 600 nm for an excitation wavelength of 380 nm. For InSb and CdS loaded TiO_2 photoanode a shift in emission towards higher wavelength side can be seen.

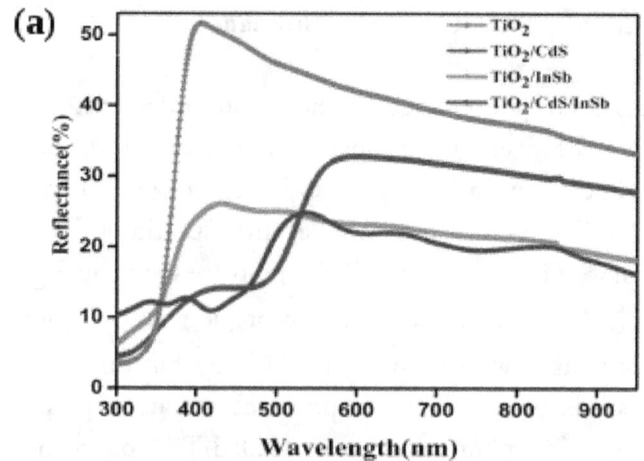

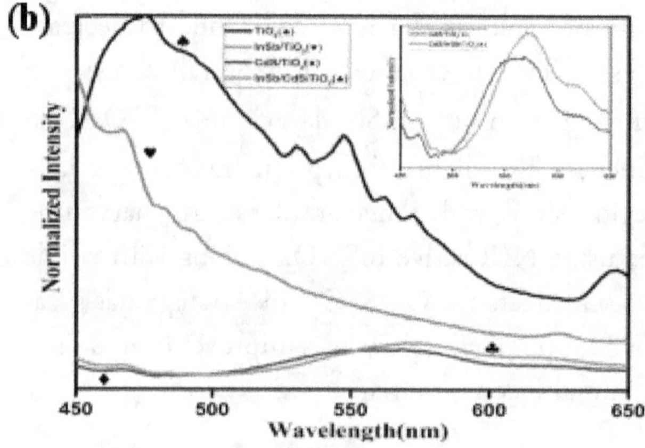

Figure 4.9 UV-Vis-NIR DRS (a) and photoluminescence spectra (b) of CdS and InSb QDs loaded photoanode

The emission at two different wavelengths were due to difference in band gap of CdS and InSb as seen by difference in absorption occurring at visible and NIR region (Kalanur *et al.* 2013, Lee *et al.* 2010; Cheng *et al.* 2017). The PL study demonstrated that the emission performance can be enhanced by co-sensitization of QDs and thus recombination of surface trapped electrons can be relatively suppressed. The PL peaks of TiO₂/CdS/InSb photoanode were relatively boarder, which may be due to the band matching. The band alignment favours the transfer of electrons from the InSb and CdS QD sensitizer layers towards photoanode layer, with reduced charge recombination

EIS Analysis

EIS analysis is used to evaluate the internal resistance and charge transfer kinetics of fabricated photoanodes. Figure 4.10 shows the EIS plots of TiO₂/CdS and TiO₂/CdS/InSb photoanodes. The plots were analysed using EC lab software with an equivalent circuit. From EIS analysis, it was clear that the TiO₂/CdS electrode shows relatively higher R_s compared to TiO₂/CdS/InSb co- sensitized photoanode.

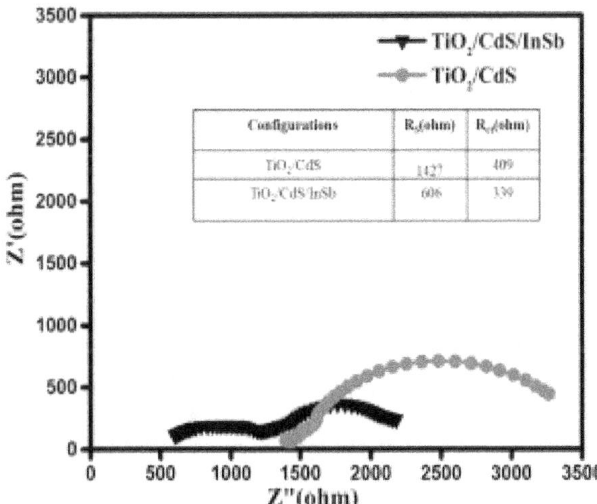

Configurations	R_s(ohm)	R_{ct}(ohm)
TiO₂/CdS	1427	409
TiO₂/CdS/InSb	606	339

Figure 4.10 Electrochemical Immepedence spectra of QDSSC cells with platinum Counter Electrode (inset table is observed values for Rc and R_{ct})

From EIS, it was also confirmed that the $TiO_2/CdS/InSb$ co-sensitized photoanode shows low R_{ct} compared to TiO_2/CdS electrode which favours for high performance of QDSSC with co-sensitized photoanodes. The inset table shows the measured series resistance (R_s) and R_{ct} of the fabricated QDSSC. The low R_s and R_{ct} of the co-sensitized photoanodes are originated from the narrow band gap InSb QDs.

J-V Analysis

The J–V characteristics of fabricated QDSSCs were analysed using a standard (AM1.5) solar simulator (Oriel Class AAA) under 1 sun illumination condition. The obtained cell parameter values are shown in Table 4.1. Figure

4.10 shows the J–V curves of bare InSb QDs, CdS QDs and InSb/CdS co- sensitized QDs based QDSSCs fabricated using TiO_2 as photoanodes and Pt and CuS as CE. As can be seen from Figure 4.11, the QDSSC with CuS CE showed relatively better performance compared to Pt CEs for all the cell configurations. This was obviously because of better compatibility of polysulphide electrolyte with CuS CE than that with Pt CE. High electrocatalytic activity of CuS electrode reduces R_{ct} at electrolyte/electrode interface and thus not only reduces internal resistance but also enhances cell performance (Li *et al.* 2014). From the J–V curves, it can be inferred that the QDSSC fabricated using InSb QDs with Pt CE have shown an effciency of 0.51

% and the cell with CuS CE shows the efficiency of 0.8 %. While J–V characteristics of CdS based QDSSCs shows an efficiency of 1.25 % for Pt and

3.52 % for CuS CE.

Co-sensitization of CdS and InSb in QDSSC with CuS CE showed an efficiency of 4.94 %, which is significantly enhanced compared to all other cell structures (Table 4.1). Further the J–V curves demonstrated the relative enhancement of J_{sc} (18.58 mA/cm$_{-2}$) and FF (0.49) on co-sensitization which may be originated from the broadened light absorption spectrum of co- sensitizers as seen from UV-Vis-NIR spectra (Figure 4.9 a). Wide solar spectrum coverage provides opportunities for more electronic excitation of

QDs at different energy levels thereby generating multiple excitons. The band matching achieved by each QD layer without spectral overlap favours electron hopping at a faster rate towards the photoanode which improved J_{sc}.

Table 4.1 Cell parameters of the QDSSCs with and without co- sensitization

Cell	Cell configuration	Voc (mV)	Jsc (mA/ cm2)	FF	Efficiency (%)
QDSSC1	TiO2/InSb /Pt	186	10.96	0.24	0.51
QDSSC2	TiO2/CdS/Pt	367	8.83	0.38	1.25
QDSSC3	TiO2/CdS/InSb/Pt	516	8.63	0.39	1.77
QDSSC4	TiO2/InSb/CuS	287	10.71	0.26	0.81
QDSSC5	TiO2/CdS/CuS	454	16.98	0.45	3.52
QDSSC6	TiO2/CdS/InSb/ CuS	533	18.58	0.49	4.94

The V_{oc} of the QDSSC has been significantly improved for the co-sensitizer compared to single sensitizer based QDSSCs. In general, V_{oc} was mainly affected by the recombination losses that can deliberately downturn excitonic life time of charge carriers. A significant improvement in V_{oc} was observed for co-sensitization of IR active InSb with visible active CdS in QDSSCs which resulted the low recombination losses. InSb QDs act as an efficient light absorber and passivating layer for SILAR deposited CdS in the photoanodes thereby decreasing the surface trap sites as seen from PL spectra in Figure 4.9 (b). In addition, the InSb layer prevents retention of electrolyte towards sensitizer and hence much better control on recombination process was resulted (Khalili *et al.* 2018). Furthermore, the TiO2/CdS/InSb photoanodes shows relatively low R_{ct} and low R_s as observed from EIS spectra in Figure

4.10 which ease the charge transport in the fabricated QDSSCs. Therefore, the InSb co-sensitized QDSSC shows high effciency of 4.94 %. The obtained efficiency was relatively higher than that of reported co-sensitized QDSSC (Lin *et al.* 2017; Hwang *et al.* 2013). The experimental results demonstrated that the co-sensitization of IR active InSb QDs with CdS was a promising approach to improve the performance of QDSSC.

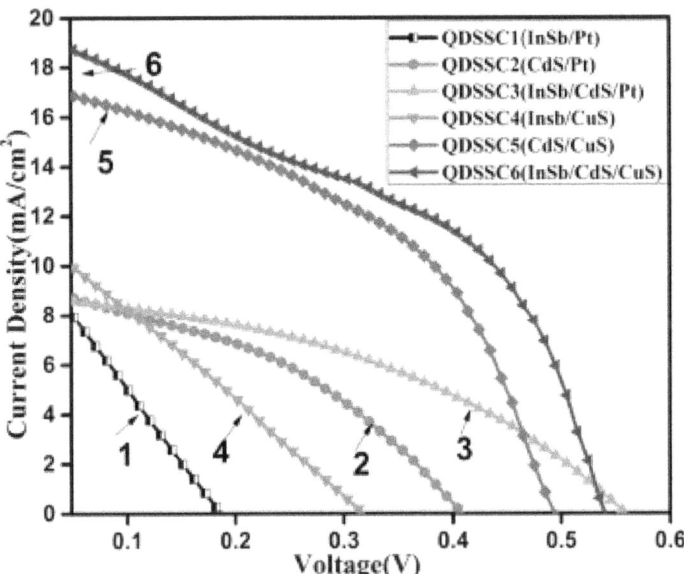

Figure 4.11 J-V characteristics of fabricated QDSSCs with different configuration

4.4 CONCLUSION

InSb QDs has been successfully synthesized by solvothermal method. The structural, morphological and optical properties of InSb QDs were investigated. To improve the efficiency of QDSSC, the IR active InSb QDs were effectively co-sensitized with CdS QDs and studied their J–V characteristics. The J–V results indicated that the InSb QD co-sensitized QDSSC shows relatively higher efficiency of 4.94 % compared to CdS QD based QDSSC (3.5 %). Moreover, the co-sensitization of InSb QDs, effectively improved the J_{sc} (18.57 mA/cm 2) and V_{oc} (533 mV) of QDSSC which resulted the high efficiency. The co-sensitization of InSb QDs improved the absorption of solar energy and suppressed the recombination losses which improved the cell performance. Therefore, the present study emphasizes the potential of co- sensitization of InSb QDs for QDSSC.

CHAPTER 5

SYNTHESIS AND CHARACTERIZATION OF GRAPHENE QUANTUM DOTS AS PASSIVATION LAYER FOR QUANTUM DOT SENSITIZED SOLAR CELLS

INTRODUCTION

Quantum Dot Sensitized Solar Cells (QDSSC) is an alternative technology for dye sensitized solar cells despite of its low efficiency (Ruhle *et al.* 2010). To improve the photovoltaic performance of QDSSC various approaches such as modifying sensitizer by various QD materials with tunable light absorption range, various electrolytes including polysulfide and cobalt complexes, and different CEs (e. g. Pt, Au, Cu_2S, carbon materials) have been attempted (Duan *et al.* 2015). Introduction of passivation layers (ZnS, SiO_2) for controlling recombination at the QD/electrolyte interface have also been investigated (Rao *et al.* 2015; Lee *et al.* 2017; Sharma *et al.* 2016; Liu *et al.* 2020). But still the conversion efficiencies of QDSSCs are limited and it is highly challenging to improve the efficiency of QDSSCs.

Graphene has been attracted great attention as a nontoxic, eco-friendly, chemically inert 2D material with high electrical and thermal conductivities (Gupta *et al.* 2011). The excellent characteristics of graphene had led to its potential applications in the fields of light-emitting diodes, photovoltaic devices, bioimaging probes, photodynamic therapy, photocatalysis and flexible touch screens (Shen *et al.* 2012). Graphene Quantum Dots (GQDs) are the one-atom-thick discrete 2D material with sp_2 carbon atoms. The quantum confinement of the electrons in GQDs gives various novel optical and electronic properties (Yan *et al.* 2019). As the valence and conduction bands (π and π^* orbitals) meet at the Dirac point, graphene is considered as a zero-bandgap material. On shrinking the three dimensions of graphene to nano regime, band gap widens up in GQDs in a size dependent manner (Yan *et al.* 2018).

By tuning the band gap optical property like photoluminescence can be induced in GQDs. Band gap modification can be done by modifying size and morphology of sp2 domains in the sp3 matrix (Sarkar *et al.* 2016b). Photoluminescence mechanism of GQDs is originated from quantum confinement effect, defect and edge effect. Moreover, both theoretical and experimental studies have shown that the structural formation and chemical feature of edges in GQDs plays an important prerequisite role in determining their electronic and magnetic properties (Kim *et al.* 2019). High value of intrinsic mobility, favourable electrical and thermal conductivities, larger quantum yield (QY) fluorescence emission, excellent photostability, good water solubility, lower value of cytotoxicity, pervasive biocompatibility are the remarkable properties of GQDs (Gao *et al.* 2019).

As of now diverse roles of GQDs in developments of photoanodes, CEs and even as supporting layer for photoanode have been studied (Kumar *et al.* 2013;Chetia *et al.* 2016). Moreover, GQDs have been also introduced as sensitizer for the QDSSC by Yang *et al* (2010). as it possesses favourable semiconducting property. These diverse applications of GQDs in QDSSC demands more research on graphene based solar cells and its hybrids in order to fully optimize the potential of this material in QDSSCs (Tian *et al.* 2018).

The use of additional passivating layer promotes light absorption and enhances charge carrier separation thereby better cell performance can be realized. More than a sensitizer material, GQDs can serve as interfacial passivating layer between sensitizer and electrolyte offering better sensitizer stability with suppressed recombination. This hybrid layered QDSSCs have superior photocurrents, cell voltages, and overall efficiencies than those of monolayer QD sensitizer based QDSSCs (Sharif *et al.* 2019; Gao *et al.* 2019; Kumar *et al.* 2019a; Majumder *et al.* 2019). As they are biocompatible, and cost effective, the GQDs might be the right choice as an alternative replacement for passivating layer along with the sensitizing materials for TiO_2 photoanodes in QDSSCs.

Several studies have been reported on controlling charge recombination in CdS based QDSSCs by using ZnS QDs as passivation layers that suppress leakage current between CdS QDs and electrolytes (Rao *et al.* 2015; Lee *et al.* 2017; Sharma *et al.* 2016; Liu *et al.* 2020; Ha *et al.* 2015). The electrical conductivity of GQDs are extremely higher than that of ZnS QDs and thereby more efficient charge separation can be achieved on replacing ZnS by GQDs. Hence strategy, linking with full-grown carbon chemistry, unfolds exciting opportunities to tune the optical and electronic properties of GQDs is needed for photovoltaics application (Roy-Mayhew *et al.* 2014; Wang *et al.* 2015d). Therefore, in this chapter, GQDs were synthesized by probe sonication assisted liquid phase exfoliation method and used as a passivating layer for CdS QDs/TiO_2 based QDSSCs.

EXPERIMENTAL

Preparation of Graphene Quantum Dots by Liquid Exfoliation

Graphite powder with high purity (99.99%) was procured from Sigma-Aldrich. N, N-dimethylformamide (DMF), ethylacetoacetate (EAA), NaOH (TCI Chemicals Ltd, India) were procured and used without further purification. Minigen poly ether sulfone syringe filters (0.22 mm) and 3 spectra/por dialysis membranes, MWCO: 3500 (spectrum laboratories) were used for the synthesis process.

200 mg of NaOH was added to 120 mL of EAA under constant stirring for 60 min. After that, 400 mg of graphite powder was added to the above solution and sonicated for 120 min using probe sonicator with 10 kHz [Sonics Vibra Cell, model no. cv 18]. Dark brown solution obtained was centrifuged at 10000 rpm to remove exfoliated graphite and larger GQDs. Further the solution was filtered using a syringe filter. The solution was purified by using a dialysis membrane (cut- off 3.5 kDa) for period of 24 h. The solution outside dialysis membrane was yellow in colour and was taken as GQDs with required size. The solution was further dried by using a rotor evaporator and obtained GQD powder was dispersed in water for further characterization (Yan *et al.* 2019; Sarkar *et al.* 2016b).

Device Fabrication

Fluorine doped tin oxide (FTO) (7 Ω resistance), (TTIP) titaniumtetraisopropoxide, cadmium chloride, chloroplatinicacidhexahydrate, zinc nitrate (Sigma Aldrich), (HNO_3) nitric acid, acetone, methanol extra pure (Merck), copper nitrate, acetic acid (Alfa aesar), high purity ethanol (HPLC grade) (Honeywell), ethylcellulose, α-terpinol, sodium sulphide (TCI chemicals) were purchased.

TiO_2 paste was deposited on the conductive side of cleaned FTO substrate by doctor blade technique with cell area of 0.16 cm2. CdS QD layer was deposited by SILAR method by soaking photoanodes in 0.06 M concentration of cadmium and sulphur precursors for 8 cycles. The number of deposition cycle and precursor concentrations were optimized for the present work (Sui *et al.* 2013; Pathan *et al.* 2004; Jun *et al.* 2014; Myeong-Soo Jeong *et al.* 2014; Yu *et al.* 2012). GQDs were sonicated for 30 min and loaded on CdS loaded TiO_2 photoanode by direct absorption method. The sensitized QD loaded photoanode was dried by purging N_2 gas. Inorder to analyse the sensitizing property of GQDs it was coated by direct absorption on to TiO_2

photoanode. ZnS was coated on the CdS/TiO_2 photoanodes by SILAR method (1 cycles) (Wang *et al.*2017; Gimenez *et al.* 2009). For preparing Pt coated CE, 20 mM of chloroplatinic hexahydrate was prepared and coated by doctor blade method on to the conducting side of FTO substrates. The prepared Pt coated CE were sintered at 450 ◦C for 30 min. CuS CEs were prepared by drop casting method using 100 µL of 0.5 M Cu $(NO_3)_2$ in methanol and Na_2S in aqueous methanol solution at 3:7 volume ratio. Drop casting was done for three times and CE was soaked with ethanol and finally dried for one hour at 120 ◦C in air (Buatong *et al.* 2017; Lee *et al.* 2008). The prepared QD loaded photoanode and CE were sandwiched together with a layer of parafilm as a spacer to fill the polysulphide electrolyte.

The optical properties of the prepared samples were characterized by UV-Vis spectrophotometer (Shimadzu UV-2450). JY Fluorolog-3-11 spectrofluorometer was used to investigate the luminescence properties of the GQD samples. The presence of functional groups in the sample was studied by FTIR (Jasco 6600) spectrophotometers. Morphological analysis was performed by high resolution scanning electron microscope with energy-dispersive x-ray analysis (HR SEM-F E I Quanta FEG 200), Photovoltaic characterization of the QDSSC was carried out using solar simulator (Oriel Class AAA) instrument under one sun illumination. EIS measurements were carried out by CHI electrochemical workstation (CHI 660C) at frequency ranging from 10 MHz to 100 KHz and impedance spectra were analysed with EC lab software.

RESULT AND DISCUSSION

Optical Studies of Graphene Quantum Dots

UV-Vis absorption spectrum of GQDs is shown in Figure. 5.1 (a). The prepared GQDs shows varying light absorption behaviors.

GQDs showed a high intense absorption peak at 250 nm and a shoulder peak with weak intensity around 340-380 nm. The absorption peak at 250 nm for all GQDs was observed because of π-π^* transition in sp2 basal plane, commonly seen for graphitic materials. The broad shoulder peak at 340-380 nm was due to transition of electron from n orbital that was induced by the weak electron-donating -OH group on GQD to the excited π^* orbital (Yan *et al.* 2018 & Sarkar *et al.* 2016). From the optical absorption data, the Tauc plot is plotted and shown in Figure

5.1 (b). The GQDs possess a band gap of 2.80 eV as seen from Tauc plot in Figure 5.1 (b).

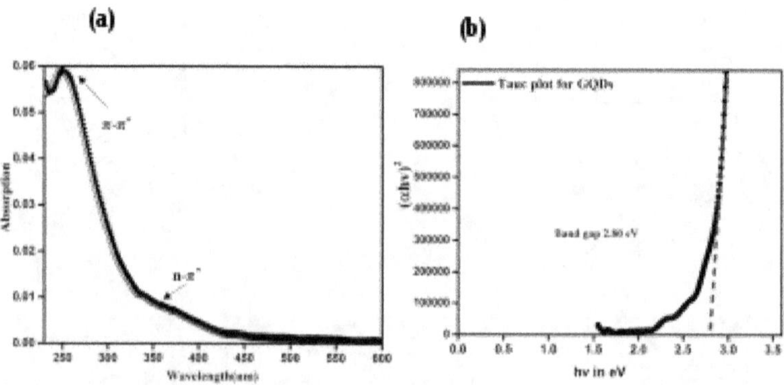

Figure 5.1 (a) UV-Vis absorption spectrum (b)Tauc plot of GQDs.

Photoluminescence and Lifetime Analysis of Graphene Quantum Dots

Figure 5.2 (a) shows the PL spectra of GQD sample at different excitation wavelength. As can be seen from Figure 5.2 a, the PL spectrum of GQDs shows two distinct peaks at around 350 and 450 nm respectively. The photoluminescence (PL) of prepared GQDs were virtually excitation- dependent originating from uneven dot size of the sample. As can be seen from Figure 5.2 (a) a broad emission peak at about 430 nm for the excitation wavelength of 300 nm. The emission peak has been shifted towards higher wavelength as the excitation wavelength increased from 300 to 550 nm. With increase in excitation wavelength an increase in defect level peak intensity of GQDs along with peak shift from 430 to 455 nm can be observed which shows the dominant defect emission from the sample. While peak corresponding to band edge emission seen to be decreased with increase in excitation wavelength. The broadened emission peaks were extended up to 550 nm which shows the possible overlapping of defect level emission with low intensity. The emission shown by GQDs was due to the combination of defect state emission (surface energy traps) and intrinsic emission due to quantum confinement effect/zigzag effect (Sarkar *et al.* 2016b, Tian *et al.* 2018). Quantum confinement effect owing to carbon core and surface state defects because of the presence of functional groups present on the surface of GQDs (Ahirwar *et al.* 2017; Sharif *et al.* 2019).

The emission spectra of the as-prepared GQDs showed excitation-dependent feature. To study excitation-dependent feature of GQDs excitation- emission spectra was analysed in detail and shown in Figure 5.2 (b). For excitation wavelength between 300 to 448 nm a clear optical response can be seen between 330 to 480 nm, which indicates the existence of a continuous emission band in the sample that can be attributed to the high degree of conjugation in GQDs, leading to the considerable overlaps of emission bands as seen from Figure 5.2 (a). A gradual decrease of band edge emission peak with only defect level emission peak can be seen beyond excitation wavelength of 360 nm. Decrease in band edge emission was due to variable band gap in GQD due to confinement effect. The excitation emission matrix of 3D- fluorescence topographical map of GQDs illustrates that no absorption or excitation occurs beyond wavelength of 450 nm. Hence it can be clearly inferred that the optical activity of GQD lies in between 280 to 480 nm with defect level emission existing from 420 to 480 nm (Yan *et al.* 2018)

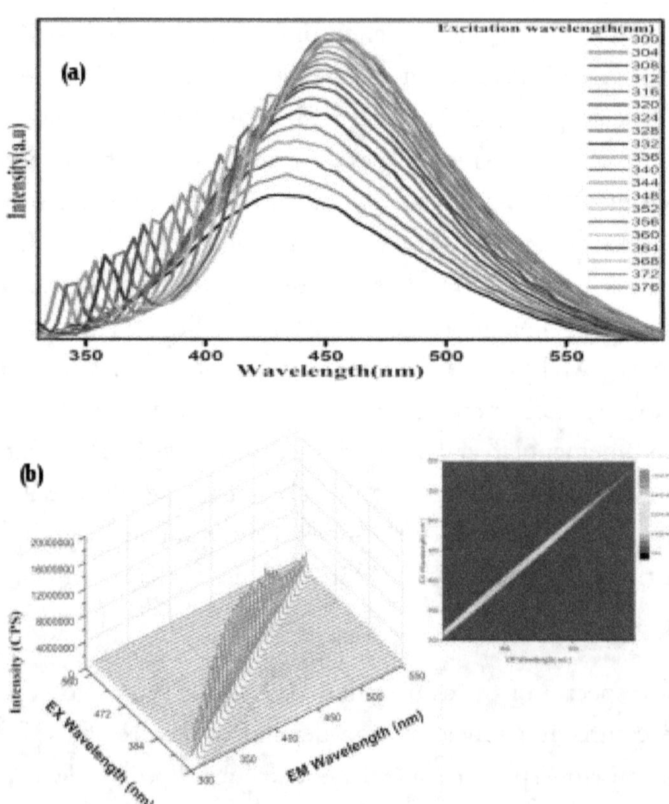

Figure 5.2 Photoluminesence spectra(a) excitation emission matrix (b) of Graphene Quantum Dots

GQDs fluorescence life time is experimentally analyzed using fluorescence life time measurement system to study suitability of GQDs for optoelectronic applications. GQDs showed nanosecond lifetime of 1.25 ns for excitation wavelength of 295 nm and emission at 433 nm as shown in the PL decay curves (Figure 5.3). GQD with lifetime in nanoscale level confirms the singlet state nature of the GQD emission (Ahirwar *et al.* 2017).

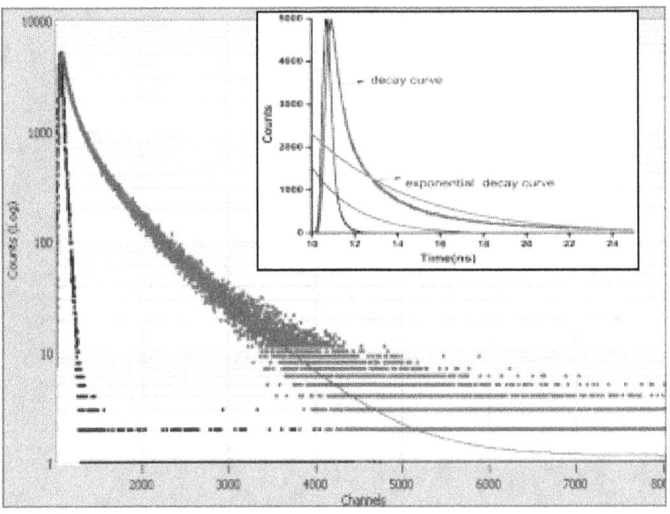

Figure 5.3 Fluorescence lifetime distribution curve (inset: exponential decay curve) of GQDs

FTIR Analysis of Graphene Quantum Dots

Figure 5.4 (a) and (b) shows the FTIR spectra of GQDs and graphite samples. From FTIR spectra of GQDs peaks corresponding to stretching frequency of O–H, C-O, and C=O groups can be identified. The broadened peak at 3327 cm-1 was assigned to the stretching vibration of O-H and band at 1359 cm-1 was due to bending vibration of O–H groups while bands at 2158 cm-1 and 729 cm-1 were due to the stretching and bending vibrations of C–H groups (Sankar *et al.*2016b). The bands around 1747 cm-1 due to the stretching of C=O groups, respectively. The peak at 1625 cm-1 was attributed to the stretching vibration of C=C (Yan *et al.* 2019). More disordered structure can be seen for as prepared GQDs than graphite powder from Figure 5.4 (a) and (b) due to presence of more oxygen containing functional groups. As a result of high probe sonication reaction of graphite, NaOH and EAA, the oxygen containing functional groups were introduced at the edges and basal planes of the graphene sheets of GQDs, FTIR spectra confirms the presence of functional groups on surface of GQDs causing surface defect and hence it resulted strong defect related emission as observed in PL spectra of GQDs (Figure 5.2a).

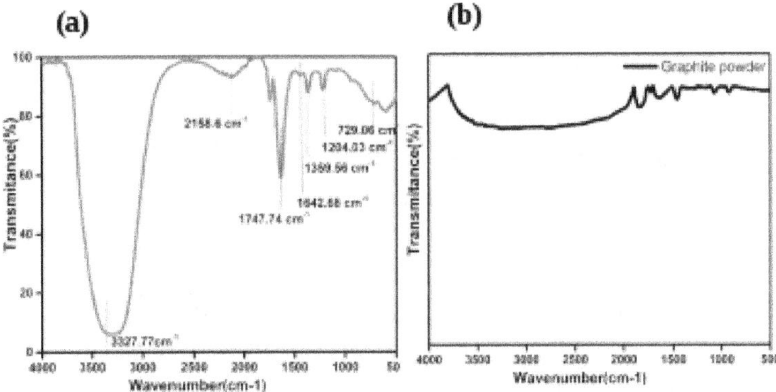

Figure 5.4 FTIR spectra of (a) GQDs (b)Graphite

Raman Analysis of Graphene Quantum Dots

Figure 5.5 shows the Raman spectra of GQDs and graphite samples. In both the spectra, two distinctive D and G bands were clearly observed. D band was disorder induced band that confirms presence of disordered SP_2 network in the sample. It was due to more oxygen containing functional groups introduced on to graphitic structure under extreme ultra sonication reaction. G band was due to C-C stretching of graphitic carbon. For the graphitic sample, the G band was obtained at 1595 cm -1 which was dominant due to graphitic nature of sample. For GQDs it can be inferred that defect level peak intensity was increased compared to graphitic peak intensity. Degree of disorder is the measure of intensity ratio of D band to G band and can be seem to be greater than 1 for GQDs. The Raman spectra confirms that more defects were present in GQDs in comparison to graphite.

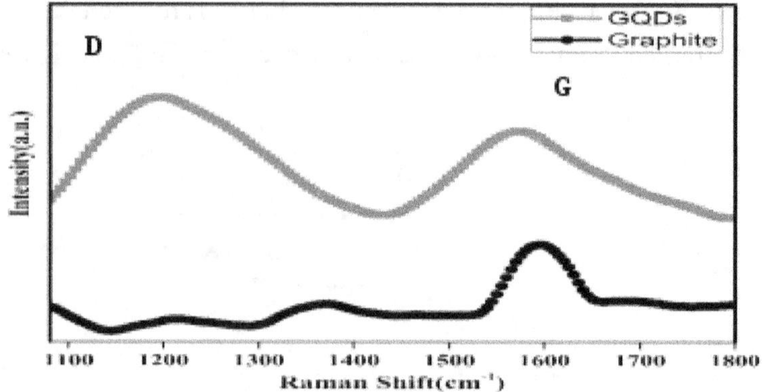

Figure 5.5 Raman spectra of GQDs and graphite

TEM analysis of Graphene Quantum Dots

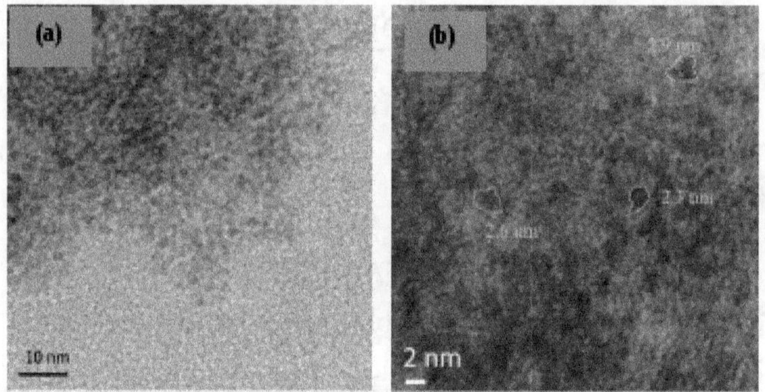

Figure 5.6 a and b: TEM images of Graphene Quantum Dots

Figure 5.6 (a) and (b) shows the HRTEM images of GQDs. From HRTEM images the formed smaller size GQD particles were confirmed to be QDs with uneven size distribution. The GQDs were irregular in shape and the sizes were less than 10 nm as can be observed from Figure 5.6 (b).

FESEM analysis of GQD loaded on TiO$_2$ Photoanode

Figure 5.7 (a) shows FESEM image of GQDs loaded on TiO$_2$ photoanode and its EDS spectrum. The FESEM images confirm the presence of smaller size GQDs on larger size TiO$_2$ and effective GQD loading by DA method can also be confirmed. The presence of C, Ti and O along with its elemental composition can be observed from EDAX table in Figure. 5.7 (b).

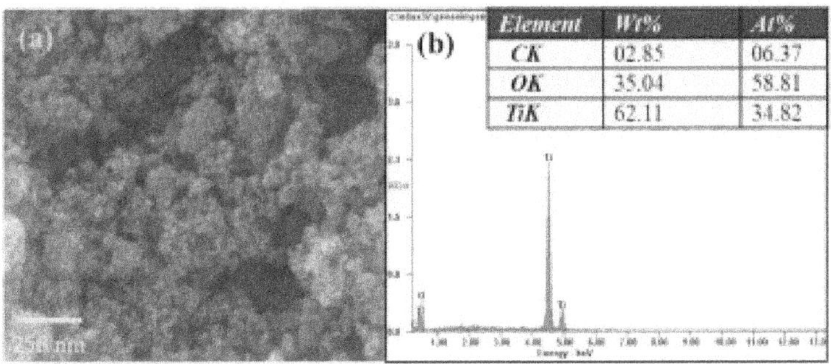

Element	Wt%	At%
CK	02.85	06.37
OK	35.04	58.81
TiK	62.11	34.82

Figure 5.7 FESEM image (a) and EDAX with elemental composition (b) of GQDs on TiO$_2$ photoanode

AFM analysis of GQDs loaded TiO$_2$ photoanode

Figure 5.8 (i) a-d shows 2D AFM images and Figure 5.8 (ii) a-d shows 3D AFM images of bare TiO$_2$ photoanode, CdS loaded TiO$_2$ photoanode, GQD passivated CdS sensitized photoanode and ZnS/GQD co-passivated CdS loaded photoanode respectively. On analyzing surface features from AFM images (Figure 5.8(a-d)) it can be observed that bare TiO$_2$ and CdS loaded photoanode showed more surface roughness compared to GQD passivated photoanode. For ZnS and GQD co-passivated CdS sensitized photoanode the surface roughness decreased as seen from Figure 5.8 (c and d). More smoothened surface can be observed for CdS loaded with ZnS and

GQDs due to the passivation of photoanode by ZnS and GQDs that reduced the surface roughness.

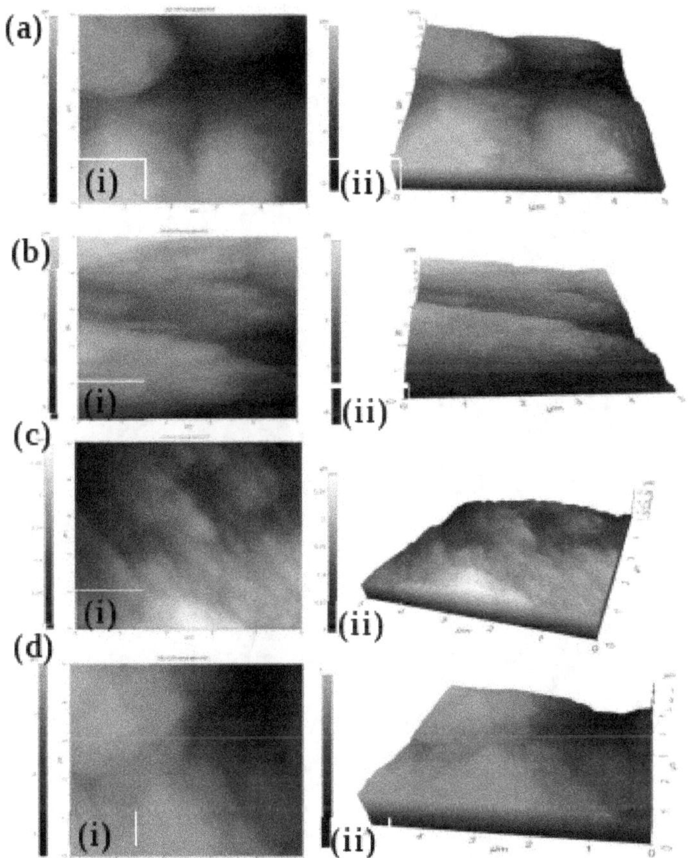

Figure 5.8 (i) 2D and (ii) 3D AFM images of (a) bare TiO₂ photoanode (b) CdS loaded TiO₂ photoanode (C9) (c) GQDs passivated CdS loaded TiO₂ photoanode (d) GQDs/ZnS co-passivated CdS loaded TiO₂ photoanode

J-V Characteristics

Figure 5.9 (a) and (b) shows the J-V characteristics of GQD passivated QDSSCs and their performance was compared with ZnS passivated QDSSC and GQD sensitized cells. Table 5.1 shows the comparison of J-V characteristics of fabricated QDSSCs. As can be seen from Table 5.1, J_{sc} (8.7 mA. cm-2) and V_{oc} (518.5 mV) of GQD passivated CdS QDSSCs were relatively higher than that of ZnS passivated QDSSC (J_{sc} = 7 mA.cm-2 and V_{oc}

=466.8 mV) with Pt CE. The J_{sc} and V_{oc} has been reasonably improved for the QDSSC when the Pt CE was replaced by CuS CE (V_{oc} =555.67 mV, J_{sc}=12.83 mA.cm-2). The obtained cell performance of GQD passivated CdS QDSSCs with CuS CEs were relatively higher than that of previous reports (Ito *et al.* 2007; Pathan *et al.* 2004; Jun *et al.* 2014). The performance of QDSSC with GQDs as sensitizer was very low compared to CdS based QDSSC. On the other hand, a significant improvement in V_{oc} has been observed in CdS based QDSSC with GQDs passivation. The introduction of GQD as a passivation layer on the surface of CdS QDs, preventing leakage current from the electrolyte to the CdS QD layer and also hindering back electron transfer from photoanode and QDs towards electrolyte. As a result, the interfacial charge recombination was effectively suppressed which resulted high V_{oc}.

High efficiency of 4.26 % was achieved for the QDSSC with double passivation layers of GQDs and ZnS. Besides retarding charge recombination at the photoelectrode/electrolyte interface, the GQD/ZnS passivation suppressed the surface defects in sensitized photoanodes and reduced the charge recombination. Therefore, double passivation gave superior cell performance with improved J_{sc} (13.15 mA.cm-2), V_{oc} (562.57 mV) and FF (0.575) compared to ZnS passivated CdS QDSSC as shown in Table 5.1.

Table 5.1 J–V characteristics of QDSSCs fabricated with different combinations

Cell configurations	V_{oc} (mV)	J_{sc} (mA/cm$_2$)	FF	Efficiency (%)
TiO$_2$/Graphene/Pt	146.78	0.503	0.26	0.02
TiO$_2$/CdS/ZnS/Pt	466.82	7.51	0.37	1.30

TiO_2/CdS/Graphene/Pt	518.55	8.78	0.35 1.63
TiO_2/CdS/ZnS/CuS	509.86	11.70	0.54 3.23
TiO_2/CdS/Graphene/CuS	555.67	12.83	0.56 4.06
TiO_2/CdS/Graphene/ZnS/CuS	562.57	13.15	0.57 4.26

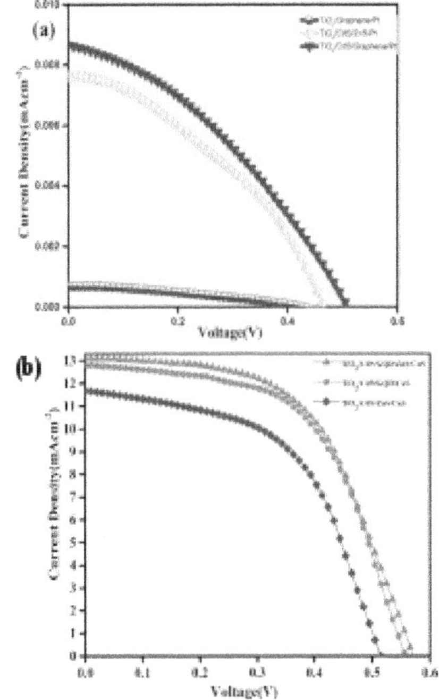

Figure 5.9 J–V Characteristics of fabricated QDSSCs (a) Pt (b) CuS CEs

EIS Characteristics

Interfacial charge transfer act as a limiting factor in enhancing overall performance of QDSSC (Liu *et al.* 2020). As ZnS was the most frequently used passivation layer material in QDSSCs, the performance of CdS QDSSCs prepared with ZnS and GQDs as passivation layers were compared. GQD synthesized by probe sonication exhibited semiconductor behaviour with a band gap of 2.80 eV from Tauc plot in Figure 5.1b which was relatively higher than the theoretical value (2.5 eV) (Sharma *et. al* 2016b) and hence seem to be a good alternate replacement of ZnS. As band gap of GQDs was smaller than ZnS (3.2 eV) and larger than CdS (2.4 eV), it was comparable to energy band distribution of CdS QDs and minimizing energy barriers at interfaces of CdS QDSSCs. As a result, much improved performance has been observed for the QDSSC with GQDs/ZnS co-passivation layers.

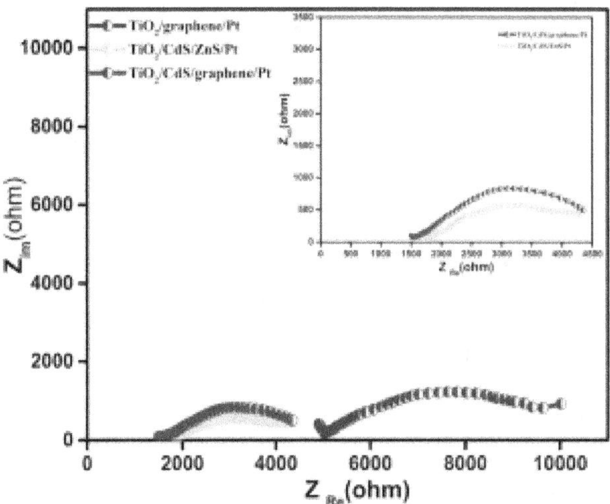

Figure 5.10 Electrochemical Impedance Spectra of assembled QDSSCs with different cell configurations

The EIS spectra of GQD passivated and ZnS passivated QDSSCs were shown in Figure. 5.10. From the EIS spectra, the equivalent R_s and R_{ct} of the QDSSC with different combinations were calculated and shown in Table

5.2. The QDSSC with GQDs passivating layer shows relatively low R_s compared to ZnS passivated QDSSC (Table 5.2). The diameter of semicircle region represents the R_{ct} which was directly proportional to charge recombination resistance at photoanode/QDs/electrolyte interface. Higher values of R_{ct} obviously rise up a challenging situation for the electrons in the photoanodes to recombine with the holes in electrolyte, leading to a reduced charge recombination. It can be inferred from EIS plot that the R_{ct} of QDSSC with GQD passivation was relatively higher (639.10 Ω) than that with ZnS passivation (516.7 Ω). Improved R_{ct} supressed the charge recombination and consequently enhanced Voc. As a result, the GQD passivated QDSSC showed relatively high Voc (518.55 mV) compared to ZnS passivated cell (466.82 mV) with Pt CE. Similar trend was observed for the QDSSC with CuS CE. Therefore, the efficiency of GQD passivated QDSSC has been improved compared to ZnS passivated QDSSC. The experimental results demonstrated that the efficiency of GQD passivated QDSSC was relatively enhanced up to

25.3 % of ZnS passivated QDSSC. Moreover, the efficiency of QDSSC has been further improved by co-passivation of GQD/ZnS.

Table 5.2 EIS characteristics of QDSSCs fabricated with different combinations.

Cell Configurations	Rs(ohm)	Rct(ohm)
TiO$_2$/Graphene/Pt	4997	403.9
TiO$_2$/CdS/ZnS/Pt	1643	516.7
TiO$_2$/CdS/Graphene/Pt	1444	639.10

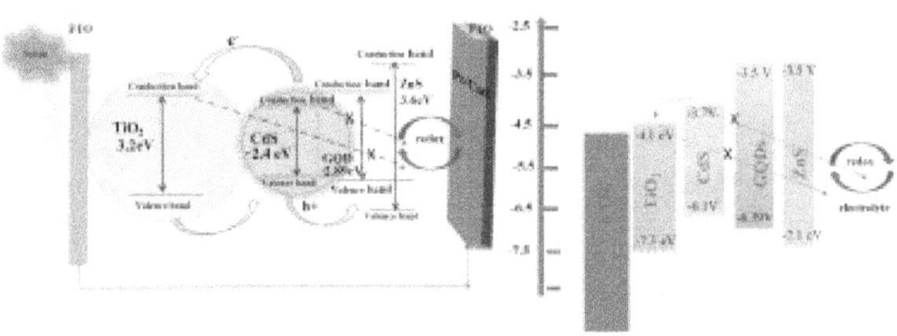

Figure 5.11 Schematic representation and energy level diagram of double passivated (ZnS/GQDs) CdS QDSSCs

Schematic diagram of energy band gap and charge transfer mechanism of TiO$_2$/CdS QDSSC with GQD/ZnS passivation layer was shown in Figure. 5.11. Incident photons with energies greater than the band gap of the CdS QDs were absorbed by sensitizer and generate excitons. The excited electrons move towards semiconductor oxide material and to external circuitry. Valence band energy level of GQDs were higher than that of CdS QDs, which can be favourable for hole transfer. (Duan *et al.* 2015; Sharma *et al.* 2016b; Liu *et al.* 2020; Luo *et al.* 2018). Simultaneously the presence of additional passivation layers of ZnS, further decreased the excitonic carrier recombination (Jiao *et al.* 2015). While the electrons flow towards conduction band of TiO_2 under the driving force acquired from the stepwise energy band structure. Accordingly, the charge carriers congregate at their respective electrodes

5.4 CONCLUSION

GQDs has been synthesised by a facile method and used as a passivating layer for CdS QDSSC. The optical and morphological properties of the GQD were analysed by UV-Vis spectroscopy and TEM analysis. The performance of GQD passivated QDSSC has been comparatively analysed with ZnS passivated QDSSC. GQD passivated QDSSC showed relatively higher efficiency (4.06 %) compared to ZnS passivated QDSSC (3.23 %) with CuS CE. The results demonstrated that GQDs could serve as an effective passivating layer which suppressed charge recombination and as a bridge inhibiting back electron from electrolyte, leading to a significantly enhanced V_{oc} (555.67 mV) compared to ZnS passivated QDSSC which results high efficiency. GQD/ZnS co-passivation further improved the cell performance by inhibiting back electron transfer from the photoanode and QDs to the electrolyte, leading to a significant decrease in interfacial charge recombination resulting high V_{oc} of

562.57 mV and high efficiency of 4.26 %.

CHAPTER 6

THE EFFECT OF MODIFIED ELECTROLYTE AND COUNTER ELECTRODE ON IMPROVING THE PERFORMANCE OF QUANTUM DOT SENSITIZED SOLAR CELL

INTRODUCTION

One of the main issues with polysulphide electrolyte used in QDSSC is its high regeneration rate of oxidized QDs that leads to low V_{oc}. Moreover, low interfacial recombination resistance at CE/electrolyte interface affects the charge transfer which in turn affect the cell performance. Hence various strategies have been tried to modify the standard polysulphide electrolyte QDSSC (Jun *et al.* 2013b; Lee *et al.* 2008). Yu *et al.* (2010) reported photovoltaic characteristics of polyacrylamide-based hydrogel modified CdS/CdSe co-sensitized QDSSC. They reported relatively high efficiency of 4

% resulted from high ionic conductivity of quasi hydrogel electrolyte.

Recently, Du *et al.* (2015) enhanced the efficiency of CdSe based QDSSC from 5.8 to 6.74 % by modifying the polysulfide electrolyte using polyethylene glycol (PEG) which hold down the charge recombination accuring at the TiO_2/QDs/electrolyte interfaces. Yu *et al.* (2017) reported the modification of polysulfide electrolyte using tetraethylorthosilicate (TEOS) as additive and studied the effect of addition of TEOS in Zn-Cu-In-Se, CdSeTe and CdSe based QDSSC systems. However, the effect of TEOS addition in polysulfide electrolyte has not yet been investigated in the CdS QDs based QDSSC. Therefore, in this chapter, polysulfide electrolyte has been modified by adding TEOS to study its effect on the performance of CdS based QDSSCs.

The TEOS modified polysulfide electrolyte at different concentration has been filled into the cell and comparatively studied its performance to determine the optimum TEOS additive concentration in polysulphide electrolyte for enhancing QDSSC performance. The photoconversion efficiency of QDSSC has significantly improved up to 5.5 % with 2.5 vol. % TEOS added polysulphide electrolyte.

EXPERIMENT Materials

Titaniumtetraisopropoxide, fluorine-doped Tin Oxide (FTO) substrate (7 Ω resistance), tetraethylorthosilicate, chloroplatinic acid (Sigma Aldrich), acetic acid, cadmium chloride, copper nitrate, methanol, acetone, nitic acid (Alfa Aesar) sodium sulphide, alpha-terpinol, ethyl cellulose, zinc acetate (TCI chemicals) ethanol (Honeywell) were used as received without further purification.

Preparation of modified electrolyte

Standard polysulfide electrolyte solution was formulated using 0.5 M Na_2S, 1 M S, and 0.2 M KCl which were mixed and ground thoroughly with the help of a mortar and pestle before dissolved into the solvent media. Electrolyte solution was prepared by dissolving the above mixture into water/methanol (7:3 volume ratio) mixed solvents (Chebrolu et al. 2019; Jeong et al. 2014). The solution was maintained in dark condition at room temperature. The solution is allowed to stir for 5 h. As prepared standard electrolyte was modified by adding different volume % of tetraethylorthosilicate (1, 1.5, 2, 2.5, 3 and 3.5 vol. %).

Preparation of TiO₂ photoanode

Titania nanoparticles prepared by hydrothermal method at 240 ⁰C for a period of 12 h was coated on cleaned FTO substrate by doctor blade method. The coating area was fixed 0.16 cm². The prepared photoanode was annealed at 450 ⁰C for 15 min in a muffle furnace and allowed to cooled down slowly to room temperature.

Table 6.1 Specification of CdS Quantum Dots loaded TiO$_2$ photoanodes

TiO$_2$ photoanode	Precursor Concentration	Number of SILAR cycle
C1	0.02 M	3
C2		6
C3		9
C4	0.04 M	3
C5		6
C6		9
C7	0.06 M	3
C8		6
C9		9

CdS QDs Loading

TiO$_2$ photoanodes were bathed in anionic and cationic precursor solutions of cadmium chloride and Na$_2$S in methanol and water at

(1:1) ratio, followed by purging with N_2 gas. Anionic and cationic precursor solutions were prepared at different concentrations (0.02 M, 0.04 M & 0.06 M) to prefigure the effect of precursor concentrations. CdS QDs were loaded on TiO_2 photoanode with each precursor concentrations by SILAR method at 3, 6 and 9 numbers of deposition cycles. Soaking time for each SILAR cycle was 30 s and the deposition was carried out at ambient temperature (~ 32 °C).

Eventually, each QD sensitized photoanode was dried by spraying nitrogen gas for 5 min. The CdS QDs loaded TiO_2 photoanodes with different precursor concentrations and cycle of depositions were named as C1 to C9 as shown in Table 6.1. ZnS passivation on to CdS loaded TiO_2 photoanode were done by soaking the electrodes into 0.1 M Zn $(OAc)_2$ methanol solution and a 0.1 M Na_2S aqueous solution. The optimal immersion time was 1 min for both solutions (Ito *et al.* 2011; Ito *et al.* 2008)

Preparation of Counter Electrode

Doctor blade method was used for preparing Pt CE, by depositing a thin layer of 20 mM H_2PtCl_6 on the conducting side of well cleaned FTO substrate followed by thermal treatment at 450 °C for a duration of 30 min. CuS CEs were prepared by SILAR and drop casting methods for comparative analysis. For SILAR deposition of, CuS CE, the cleaned FTO substrates were immersed in 1M Cu $(NO_3)_2$ solution and Na_2S solution for 8 cycles. The SILAR deposited substrates were dried at 80 °C for 60 min (Balis *et al.* 2013). For CuS coating by drop casting, 100 µL of 0.5 M Cu $(NO_3)_2$ in methanol solution was dropped on to FTO.100 µL of Na_2S in mixture of water and methanol (3:7) was dropped over the initial layer. CuS formation was confirmed from the resulted immediate colour change from blue to brown and finally to black (Liu *et al.* 2019; Han *et al.* 2017). This procedure was repeated for three times followed by rinsing with ethanol. The films were dried at 120 °C for 60 min.

QDSSC Fabrication

QDSSCs were fabricated by sandwiching the CdS QDs loaded TiO_2 photoanodes and CuS CEs using binder clips. The space between electrodes was separated by parafilm film of 60 µm thickness. The TEOS modified polysulphide electrolyte was loaded into sandwiched cell region of two electrodes through capillary action. The cells were fabricated with different configurations for optimizing the cell structure with high performance.

RESULT AND DISCUSSION

Structural analysis of photoanode and counter electrode

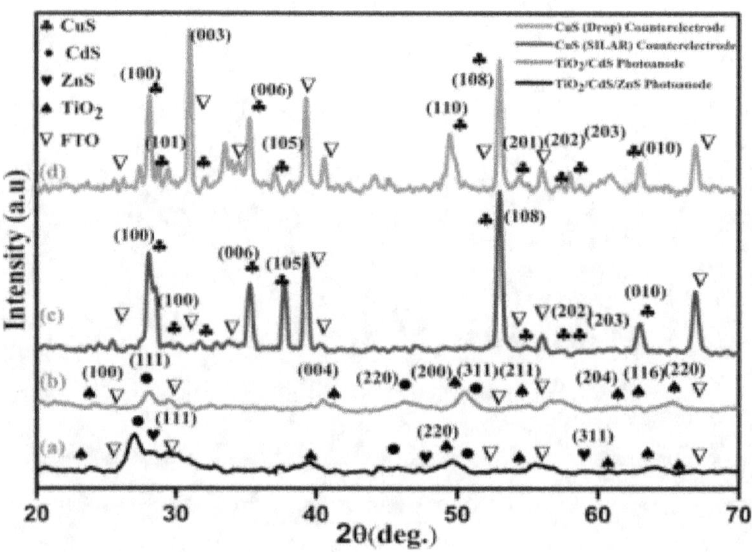

Figure 6.1 XRD diffraction patterns of (a) ZnS passivated CdS (C9) on TiO₂ photoanode (b) CdS (C9) on TiO₂ photoanode (c) CuS deposited by SILAR method on FTO (iv) CuS deposited by drop casting on FTO

Figure 6.1 (a) and (b) show the X-ray diffraction patterns of ZnS passivated and unpassivated (C9) CdS loaded TiO₂ photoanodes. The diffraction peaks of TiO₂ matches with JCPDS card 21–1272 confirming pure anatase phase of TiO₂ and CdS related diffraction peaks were well matched with JCPDS card no: 10-454. Additionally, ZnS peaks can be observed for photoanode with passivation (Figure 6.1 a) which confirmed effective passivation of ZnS on CdS sensitized photoanode. Figure 6.1 (c) and (d) shows the X-ray diffraction patterns of CE of CuS on FTO substrate prepared by SILAR and drop casting method. The diffracted peaks correspond to hexagonal CuS were well matched with JCPDS card No. 79-2321. Further, the diffraction peaks related to FTO substrate were observed for both photoanodes and CEs.

Morphological Analysis of Photoanode and Counter Electrode

Figure 6.2 a (i, ii, iii) shows the SEM images of C7, C8 and C9 samples. TiO_2 with larger grain size can be seen in the SEM image of C7 (Figure 6.2 a (i)) which shows that CdS QDs were not completely covered on the photoanode surface. While the SEM image of C9 samples (Figure 6.2 a (ii)) shows the complete coverage of ultra-small size CdS QDs on the TiO_2 photoanode surface due to the effective loading of QDs after 9 SILAR cycles of QD deposition.

SEM images confirms that C9 was an optimized photoanode for fabrication of QDSSCs and therefore further analysis was carried out for C9 photoanode. Figure 6.2 b (i & ii) shows FESEM image of CuS CE prepared by SILAR and drop casting method. Rod like morphology was obtained for the SILAR deposited CE whereas flower like morphology were observed for drop casted CuS CE.

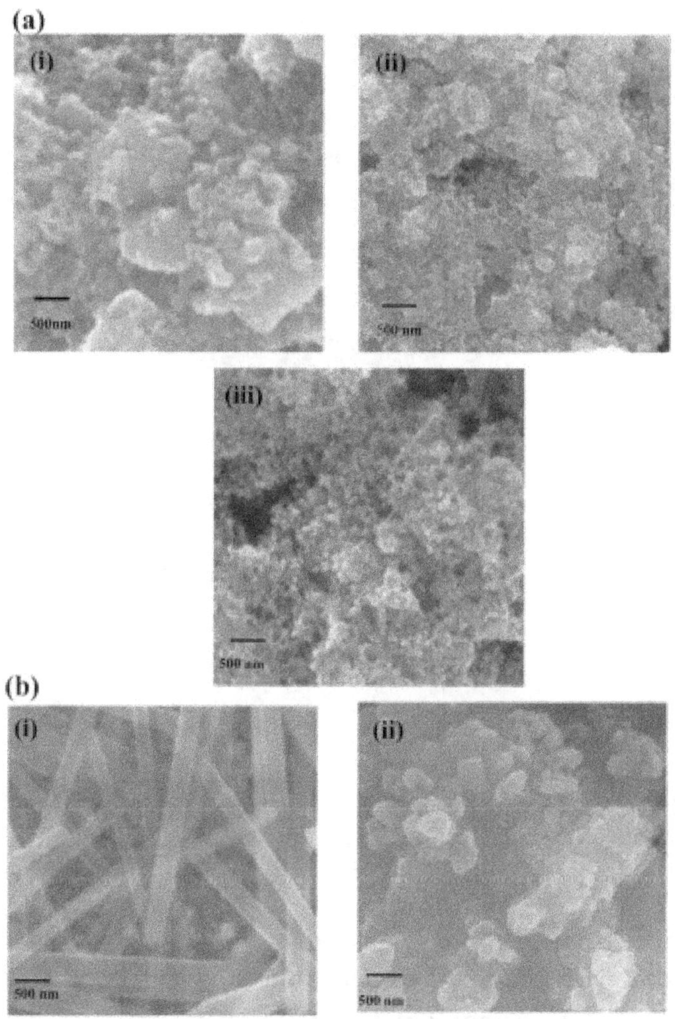

Figure 6.2 FESEM images of (a) CdS QDs loaded photoanodes (i) C7

(ii) C8 (iii) C9 (b) CuS Counter electrodes prepared by (i)SILAR (ii)Drop casting methods

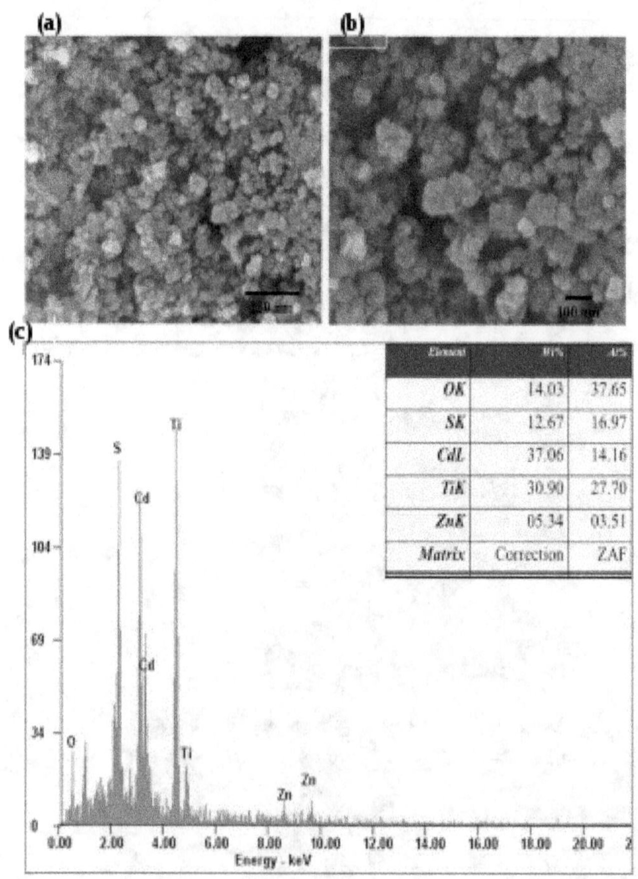

Element	Wt%	At%
OK	14.03	37.65
SK	12.67	16.97
CdL	37.06	14.16
TiK	30.90	27.70
ZnK	05.34	03.51
Matrix	Correction	ZAF

Figure 6.3 Top view of SEM mages (a and b) energy dispersive X-ray spectra with elemental composition (c)of ZnS passivated CdS (C9) loaded TiO_2 photoanodes

Figure 6. 3 (a) and (b) show SEM images of CdS loaded photoanode (C9) with ZnS passivation. The top view of SEM image shows the homogeneous distribution of CdS QDs on the surface of photoanode. Figure

(c) shows the EDAX spectrum of ZnS passivated CdS QDs loaded photoanode which confirms the presence ZnS over CdS loaded TiO_2 photoanode.

Raman Spectra of CdS loaded TiO_2 photoanode

Figure 6.4 show the Raman spectra of ZnS passivated and unpassivated CdS QDs loaded photoanodes (C9). From Figure 6.4, peaks observed at 146, 398, 517 and 638 cm-₁ confirm the anatase phase of TiO_2 which was coated over the FTO as photoanode material. ZnS passivation causes a decrease in intensity of the Raman peaks for CdS QDs loaded photoanode. Additionally, a low intensity peak was observed at about 330 cm-₁ related to CdS which was slightly shifted in the ZnS passivated sample.

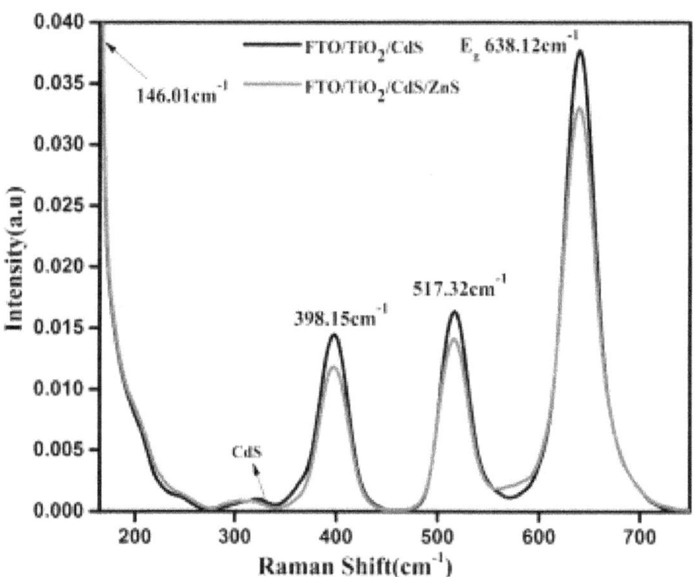

Figure 6.4 Raman spectra of ZnS passivated and unpassivated CdS loaded TiO_2 photoanodes

Optical Analysis of Photoanodes

Figure 6.5 shows UV-Vis DRS spectra of CdS QDs loaded TiO_2 photoanodes prepared with different numbers of SILAR cycles (C1 to C9).

The DRS spectra of C1, C2, C3 samples show that the reflectance onset increased from 450 to 500 nm, when SILAR cycles increased from 3 to 9 at same precursor concentration. Similar effect is observed for the other precursor concentrations of 0.04 and 0.06 M. From the DRS analysis, an obvious shift towards visible region is observed when the number of SILAR cycles increased to 9 at all precursor concentrations which confirms the effective loading of CdS QDs on TiO_2 photoanode. Moreover, low reflectivity can be seen in the samples with 9 SILAR cycles due to high absorption by QDs. In particular, maximum red shift can be observed for C9 photoanode with 0.06 M of precursor concentration and 9 SILAR cycles.

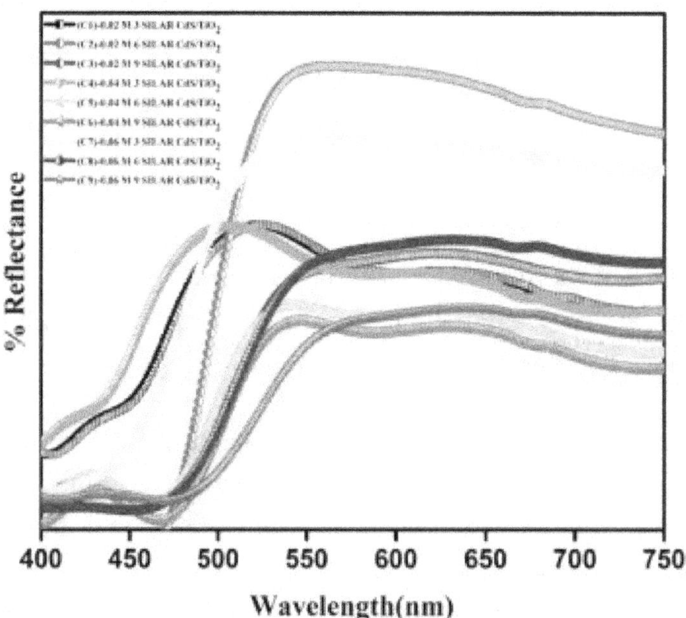

Figure 6.5 UV-Vis DRS spectra of CdS QDs loaded TiO_2 photoanodes

Photoluminescence Analysis

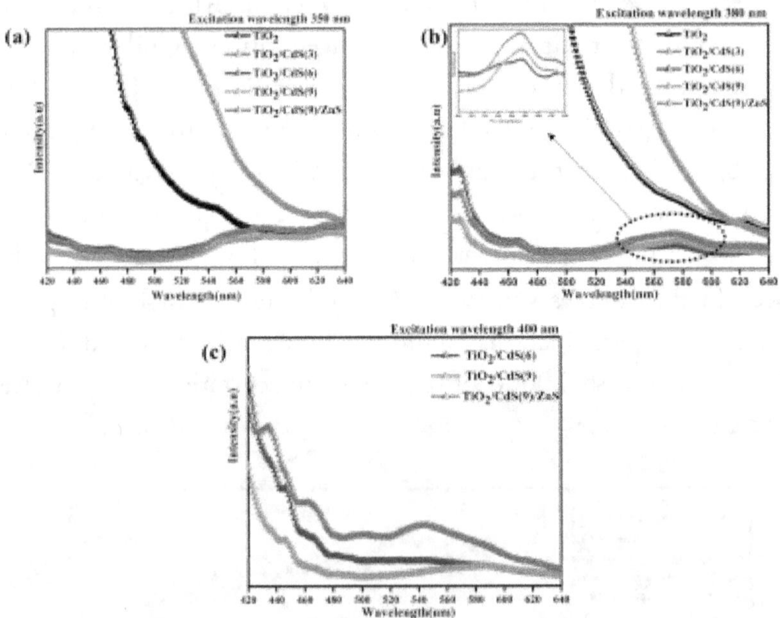

Figure 6.6 Photoluminescence spectra of TiO₂, TiO₂/CdS (3), TiO₂/CdS (6), TiO₂/CdS (9), TiO₂/CdS (9)/ZnS photoanodes for excitation wavelength (a) 350 nm (b) 380 nm (c) 400 nm

To evaluate the effect of ZnS as passivating layer, the photoluminescence (PL) spectra are recorded for different excitation wavelength of 350 nm, 380 nm and 400 nm and shown in Figure 6.6 (a-c). In general, the PL emission was the result of recombination of photogenerated electrons and holes. Hence, a weak emission was an evidence of decreased charge recombination with longer excitonic lifetime of the carriers. From Figure 6.6 (a-c), the PL emission for bare TiO_2 photoanodes, CdS QDs loaded TiO_2 photoanodes (C7, C8 and C9) and ZnS passivated (C9) photoanode can be observed near visible region around 550 nm. The emission peaks related to CdS QDs were observed in the PL spectra for the excitation wavelength of 380 and 400 nm (Figure 6.6 a), whereas those peaks were not observed in the spectra with excitation wavelength of 350 nm. The inset in Figure 6.6 (b) clearly shows the CdS QDs related emission peaks in the visible region and the intensity of the peak varied with SILAR cycle of deposition. Moreover, the peak was relatively blue shifted towards lower wavelength in the ZnS passivated C9 photoanode due to wide band gap of ZnS, which confirms the effective passivation of ZnS on the QDs surface (Figure 6.6 c).

JV and EIS Analysis of QDSSCs

Figure 6.7 (a) show the J-V curves of QDSSCs fabricated using modified electrolyte with different vol. % of TEOS and Pt CE. From the J-V curves, various photovoltaic parameters (including PCE, FF, V_{oc} and J_{sc}) were extracted and tabulated in Table 6.2. From the J-V curves, it was found that the efficiency of QDSSC increased to 1.8 % when TEOS vol. % increased up to

2.5 % in electrolyte. Figure 6.7 (b) shows the variations of average V_{oc} and J_{sc} of QDSSC as a function of TEOS concentration in the modified electrolyte. Both J_{sc} and V_{oc} of QDSSC increased reasonably with increase in the TEOS content in the polysulphide electrolyte and approached the peak values for 2.5 vol. % of TEOS. When the TEOS content increased beyond 2.5 vol. %, J_{sc} and V_{oc} were started to decline gradually and thereby the efficiency of cells decreased. Figure 6.7 (c) shows the variation of efficiency and FF of QDSSC as a function of TEOS concentration in the modified electrolyte. At higher concentration of TEOS, silicates were accumulated in the electrolyte which affected the redox potential thereby J_{sc} and efficiency decreased (Yu *et al.* 2017).

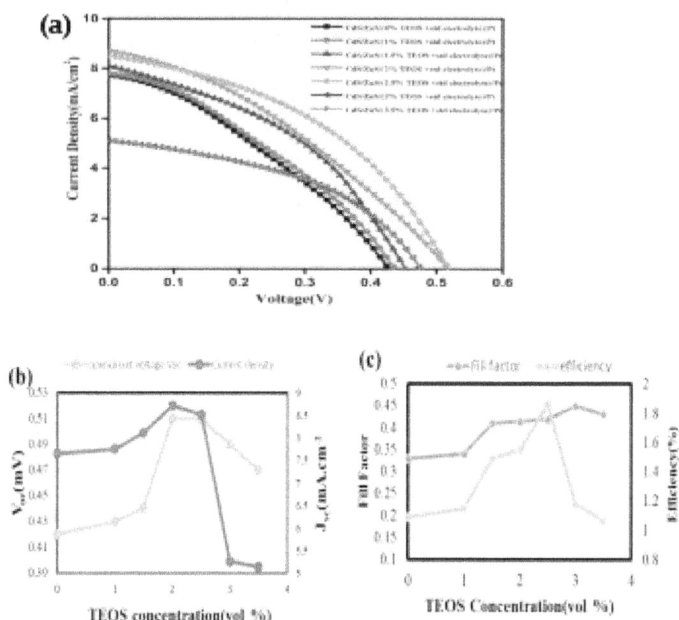

Figure 6.7 J–V curves (a) of CdS QDSSCs corresponding to modified polysulfide electrolyte with various concentrations of TEOS

(b) the dependence of average V_{oc} and J_{sc} on the TEOS concentration in the modified electrolyte (c) the dependence of the average efficiency and FF on the concentration of TEOS in the modified electrolyte

From the Figure 6.7 (a-c) 2.5 vol. % of TEOS is considered as an optimized additive concentration for modifying polysulphide electrolyte and used the same for further study. Further in order to study the role of in-situ passivation of TEOS in QDSSC, J-V and EIS analysis without ZnS passivation and shown in Figure 6.8 (a, b). The corresponding cell parameters were tabulated in Table 6.3. EIS analysis revealed high R_{ct} in the cell with TEOS modified electrolyte resulting high electron life time and hence increased charge collection efficiency. The addition of TEOS in polysulphide electrolyte resulted the in-situ passivation, which improved efficiency of QDSSC from 1.1

% to 1.5 %.

Table 6.2 J-V data of QDSSC with different vol% of TEOS added polysulphide electrolyte and platinum Counter Electrode

Cell Configurations	V_{OC} (V)	J_{SC} (mA/cm$_2$)	FF	Effic (%)
TiO$_2$/CdS/ZnS/0% TEOS+ std electrolyte / Pt	0.42	7.67	0.335	
TiO$_2$/CdS/ZnS/l% TEOS +std electrolyte/Pt	0.43	7.74	0.344	
TiO$_2$/CdS/ZnS/1.5% TEOS + std electrolyte/Pt	0.44	8.10	0.41	
TiO$_2$/CdS/ZnS/2% TEOS +std electrolyte/Pt	0.51	8.72	0.415	
TiO$_2$/CdS/ZnS/2.5% TEOS + std electrolyte/Pt	0.51	8.52	0.422	
TiO$_2$/CdS/ZnS/3% TEOS +std electrolyte/Pt	0.49	5.26	0.454	
TiO$_2$/CdS/ZnS/3.5% TEOS +std electrolyte/Pt	0.47	5.14	0.439	

Table 6.3 J-V characteristics and EIS parameters of CdS based QDSSCs using Pt CE with TEOS free and TEOS modified electrolytes

Cell Configuration	Voc (V)	Jsc (mA/ cm$_2$)	FF	Efficiency (%)	Rs (Q)	R
TiO$_2$/CdS/std electrolyte/Pt	0.358	8.08	0.39	1.14	1407	29
TiO$_2$/CdS/std electrolyte+2.5% TEOS/ Pt	0.435	8.29	0.41	1.50	1168	45

In order to further improve the efficiency, QDSSCs were fabricated by replacing Pt CE with CuS CE. The cells were fabricated using CuS CEs prepared by SILAR and drop casting methods. Moreover, to study the effect of modified electrolytes, the CuS CEs based cells were fabricated with and without TEOS in the electrolyte.

Figure 6.9 (a) show the J-V curves of CuS CE based QDSSCs with standard electrolyte and 2.5 vol. % TEOS modified electrolytes. The obtained efficiency (η), V_{oc}, J_{sc}, and FF are summarized in Table 6.4. It can be observed from Figure 6.9, that all the QDSSCs exhibited relatively higher performance with CuS CE. Moreover, the J_{sc} of the cells were one order increased compared to the cells with Pt as CE (Table 6.2) which resulted high conversion efficiency. Furthermore, the J_{sc} of cells with drop casted CuS CE significantly improved compared to the cells with SILAR CuS CE. As a result, the cells with drop casted CuS CE based QDSSC with TEOS modified electrolyte exhibited high efficiency of 5.5 % compared to that of SILAR CdS based QDSSC (3.41 %).

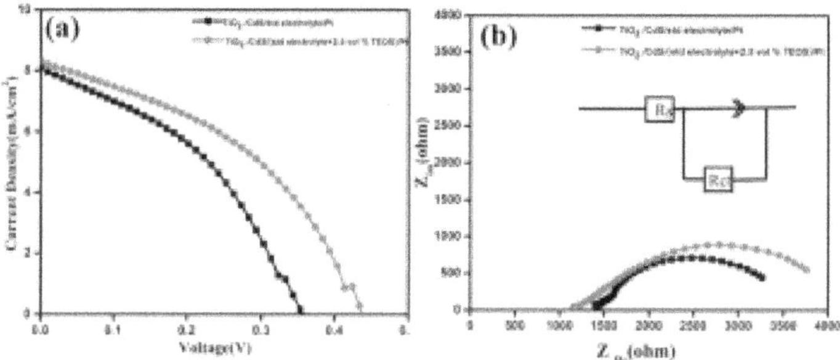

Figure 6.8 J-V characteristics(a) Electrochemical Immpedence spectra (b)of CdS based QDSSCs using Pt CE with TEOS free and TEOS modified electrolytes (inset of b EIS fitting equivalent circuit)

Figure 6.9 (b) shows the EIS spectra of cells with drop casted CuS and SILAR CuS CEs. From EIS spectra, it was clear that R_{ct} and equivalent R_s of cells with drop casted CuS CE were relatively lower than that of cells with SILAR CuS CE as the diameter of the semicircle was decreased for drop casted CuS CE based QDSSCs. Due to low internal resistance, the QDSSC with drop casted CuS

CE shows relatively high efficiency compared to SILAR CuS based QDSSC (Kalanur *et al.* 2013 & Buatong *et al.* 2017)

Table 6.4 J-V data and EIS parameters of CdS QDSSC with CuS (SILAR), CuS (Drop casted) CE with TEOS free and TEOS modified electrolytes

Cell Configuration	Voc (V)	Jsc (mA / cm2)	FF	Efficiency (%)
TiO2/CdS/ZnS/std electrolyte / CuS (SILAR)	0.480	13.14	0.488	3.08
TiO2/CdS/ZnS/ stdelectrolyte+2.5 %TEOS /CuS (SILAR)	0.480	13.69	0.519	3.41
TiO2/CdS/ZnS/std electrolyte / CuS(DROP)	0.537	12.11	0.557	3.62
TiO2/CdS/ZnS/std electiolyte+2.5 % TEOS / CuS(DROP)	0.543	19.33	0.523	5.5

The cell fabricated using drop casted CuS CE and TEOS modified electrolyte shows the best performance with V_{oc} 0.543 V, J_{sc} 19.33 mA/cm2, FF

0.523 and efficiency of 5.5 %. The efficiency was far better when compared to the performance of Pt CE based QDSSC which shows the efficiency of 1.86 %. The poor performance of QDSSC with Pt CE is mainly attributed to its affinity towards S_2- ions which slow down the catalytic activity of Pt and rapidly decreases the charge transfer rate (Faber *et al.* 2013; Wang *et al.* 2015a). Due to this fact, Pt has poor reduction rate of S_{n2}-, resulting in depletion of S_2- in the photoanode area and reduces QD regeneration rate. Whereas the QDSSCs with CuS CEs prepared by both the methods shows relatively higher performance due to low R_{CT} and R_s as observed in EIS spectra (Figure 6.9 b). The excess adsorption of S onto the Pt CE surface by chemisorption leads to a phenomenon called 'catalytic poisoning' resulting in higher resistance (Duan *et al.* 2015; Kalanur *et al.* 2013; Kumar *et al.* 2019b). Therefore, electrocatalytic process was limited by the higher internal resistance of Pt which makes CuS as an inevitable alternate CE for QDSSCs.

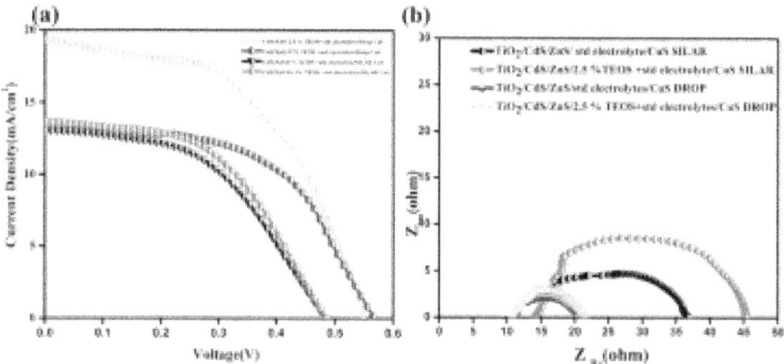

Figure 6.9 J-V characteristics (a) of CdS based QDSSCs with CuS (SILAR) and CuS (Drop casted) CEs Electrochemical immpedence spectra (b) of CdS based QDSSC with CuS (SILAR) and CuS (Drop casted) CEs

The EIS analysis of passivation free QDSSC with TEOS modified electrolyte and Pt CE throw more light on the role of TEOS in improving the cell performance. Figure 6.8 (b) shows the EIS spectra of the symmetric cells with and without TEOS in the frequency range of 0.1 Hz to 500 KHz. As can be seen from Figure 6.8 (b) & 6.9 (b), the diameter of the semicircle increased for the cell with TEOS modified electrolyte compared to cell with TEOS free polysulfide electrolyte. The larger diameter of semicircle refers the higher R$_{CT}$ which results less recombination at the electrolyte and photoanode interfaces thereby improves the performance of the cell. When TEOS added into the polysulfide electrolyte solution, it may be transformed into silica or silicon hydroxide by means of hydrolysis reaction in the strong base electrolyte solution (Yu *et al.* 2017). These particles adsorbed on the surface of CdS QDs through the hydroxyl groups and acted as a passivating layer, which suppressed the recombination process (Du *et al.* 2015). The larger diameter of EIS semicircle further confirms the surface passivation of QDs with high R_{CT} and thereby suppress the recombination of charge carriers in the presence of TEOS.

Moreover, the replacement of Pt by CuS CE reduces R_s which results relatively high performance of the QDSSCs (Kalanur *et al.* 2013 & Kim *et al.* 2017).

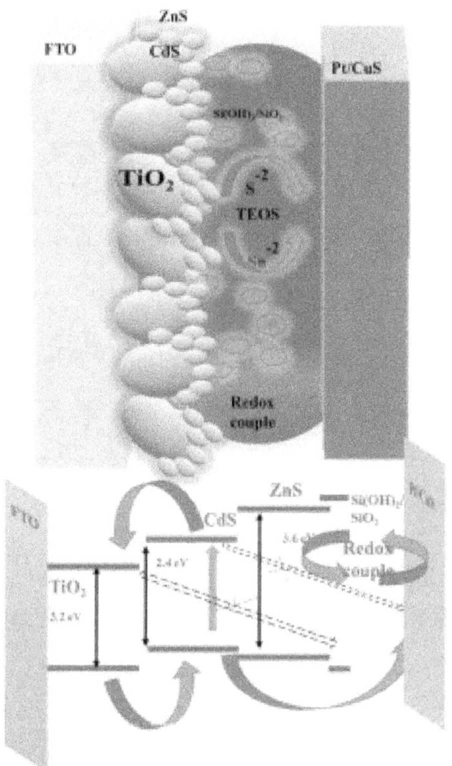

Figure 6.10 Schematic diagram for electron transfer mechanism of ZnS passivated CdS sensitized QDSSC with TEOS additive added polysulphide electrolyte

Further the schematic representation of QDSSC configuration with TEOS modified electrolyte is shown in Figure 6.10. The suppression of back electron transfer along with improved passivation resulted from insitu passivation of silica from TEOS additive greatly enhanced the overall cell performance. Therefore, it can be summarized that the QDSSC with TEOS modified electrolyte and drop casted CuS CE shows the highest efficiency of

5.5 %. The experimental results demonstrated the feasibility of further improvement in photo conversion efficiency of QDSSC by modifying the electrolyte and CE which paves an elegant way for QDSSCs to gain upcoming photovoltaic market.

6.4 CONCLUSION

QDSSCs were fabricated by modifying the electrolyte and CEs. The effective loading of QDs onto the TiO_2 photoanodes was confirmed by UV- DRS analysis. The J-V characteristics of CdS based QDSSC with optimum photoanode along with passivating layer were studied by adding different vol.

% of TEOS into polysulphide electrolyte and the cells with 2.5 vol. % of TEOS have shown an improved efficiency of 1.6 % with Pt CE. Further, on replacing Pt CE by CuS, the efficiency of QDSSC has improved significantly to 5.5 %. The role of TEOS on improving the photovoltaic performance of QDSSC was explored by EIS analysis. The addition of TEOS effectively improves the passivation of QDs thereby suppress the recombination of charge carriers which resulted high efficiency of QDSSCs.

CHAPTER 7

SUMMARY AND SUGGESTIONS FOR FUTURE WORK

SUMMARY OF PRESENT WORK

The photoanode, sensitizer, electrolyte and CE of QDSSCs have been modified and their impact on the performance of QDSSC has been systematically investigated. The experimental results demonstrated that the QDSSC with TEOS modified electrolyte and CuS CE shows the highest efficiency of 5.5 %. TiO_2 nanostructures were synthesized by sol gel and hydrothermal methods. In order to study the effect of morphology on electrical properties of TiO_2 nanostructures were synthesized by varying hydrothermal growth period as 12, 24 and 48 h for a fixed temperature of 180 ·C. The study explored the influence of synthesis conditions and post treatment process on morphology, optical and electrical characteristics of TiO_2 nanostructures. Morphological variation at different synthesis method were clearly observed in the SEM images. Spherical TiO_2 nanoparticles were formed in sol gel synthesised sample, while for hydrothermally grown 12 h sample shows sheet like 2D morphology of TiO_2 nanostructures.

On increasing hydrothermal growth period to 24 and 48 h, sheet like morphology was changed into tube like morphology. Further the effect of post treatment conditions like freeze dried at -80 ·C, dried at 100 ·C and calcined at 450 ·C were analysed on the structural, morphological, electrical and optical properties of TiO_2 nanostructures. Further, the experimental results from optical and Hall measurements demonstrated that the morphology of nanostructures has great influence on optical and electrical properties of photoanode. The hydrothermal 24 h TiO_2 nanotube exhibit low electrical resistivity and hence suitable material for photoanode of QDSSCs.

A. indica was studied as an environment friendly and cost effective nontoxic stabilizing agent for synthesizing green CdS QDs. Key research challenges like toxicity of chalcogenides were effectively controlled by green synthesis approach. The experimental results of hemolysis and cytotoxicity studies demonstrated that green CdS QDs was non-toxic compared to CdS prepared by conventional chemical method. Also, J-V studies of QDSSC fabricated with green CdS revealed that the green CdS was favourable sensitizer for QDSSC which exhibited J_{sc} of 10.61 mA/cm_2 as a proof-of-concept. Still SILAR CdS based QDSSC showed lower J_{sc} than green CdS based QDSSC. However green CdS showed lower toxicity possibly due to effective encapsulation of A. indica that can act as a stabilizing agent during the synthesis of green CdS. Additionally, ZnS passivation on green CdS suppressed the recombination and further improved the V_{oc} and efficiency of QDSSC.

InSb QDs were successfully synthesized by solvothermal method. The size of prepared InSb QDs was less than that of its excitonic Bohr radius and hence shows strong quantum confinement which makes them favourable for sensitizing applications. The structural, morphological and optical properties of InSb QDs were investigated by XRD, SEM, HRTEM, UV-Vis- NIR and PL analysis. Studies revealed that optical activity of InSb QDs were favourable as a sensitizer for QDSSC. The sensitizing properties of InSb QDs have been analysed by fabricating QDSSC and studying its characteristics. The QDSSC with InSb QDs and CuS CE has shown a photoconversion efficiency of 0.8 %. A significant improvement in efficiency was observed for QDSSC fabricated by co-sensitizing

InSb with CdS QDs. The J-V characteristic results revealed that InSb/CdS QDs co-sensitized QDSSC showed relatively high efficiency of 4.94 % compared to CdS QD based QDSSC (3.5 %). Co- sensitization of IR active InSb QDs with visible active CdS QDs in QDSSC improved the light absorption and also suppressed the recombination losses, which resulted high efficiency. The experimental results demonstrated the potential ability of InSb QDs as a sensitizer for enhancing efficiency of QDSSCs.

GQDs were synthesised by a liquid exfoliation method. As prepared GQDs were in size less than 10 nm as observed from HRTEM analysis. The GQDs showed optical absorption at 250 nm due to n-n* and a shoulder peak near 350-380 nm due to n- n* transition. Further PL study also confirms the optical activity of GQDs due to intrinsic emission and surface states defect related emission. Further analysis of GQD were done by Raman and FTIR analysis in comparison with graphite. GQDs were loaded on CdS QDSSC as a passivating layer by direct absorption method. Loading of GQDs on to CdS photoanode were confirmed by SEM and EDAX analysis. Topological surface analysis of AFM confirms good passivation of GQDs on CdS QDSSCs. GQDs passivated CdS QDSSC exhibited relatively high efficiency (4.06 %) compared to ZnS passivated QDSSC (3.23 %). GQDs served as an effective passivation layer that suppressed charge recombination which occurrs at QD/electrolyte interface of QDSSC, resulted significantly enhanced V_{oc} (555.67 mV) compared to ZnS passivated QDSSC. GQD/ZnS co-passivation further improved the QDSSC performance by inhibiting back electron transfer from the photoanode and QDs to the electrolyte, leading to a significant decrease in interfacial charge recombination resulting high V_{oc} of 562.57 mV and improved efficiency upto 4.26 %.

TiO_2 based photoanode and CuS based CE were prepared at optimized experimental conditions. QDSSCs were fabricated by modifying the polysulphide electrolyte with TEOS as additive at different vol %. The effective loading of QDs onto the TiO_2 photoanodes were initially confirmed by UV- DRS analysis. The effect of different precursor concentrations and the different number of SILAR cycle of CdS deposition on the optical properties of TiO_2 photoanode were studied by UV-Vis DRS spectra. 0.06 M precursor concentration and 9 SILAR cycle of deposition were found as optimized CdS loading on photoanode from SEM and optical analysis. The J-V characteristics of CdS based QDSSC with optimum photoanode along with passivating layer were studied with TEOS modified polysulphide electrolyte. The QDSSC with

2.5 vol. % of TEOS modified polysulphide electrolyte showed an improved efficiency of 1.6 % with Pt CE. On replacing Pt CE by CuS, the efficiency of QDSSC further improved to 5.5 %. The in-situ passivation of silica particles from TEOS additive in the polysulphide electrolyte effectively improved the QDSSC performance. The improvement in R_{ct} for QDSSC fabricated with TEOS modified electrolyte was observed in the EIS analysis. An improved R_{ct} enhanced charge collection efficiency which in turn improved electron lifetime and further improved the overall QDSSC performance.

SUGGESTIONS FOR FUTURE WORK

The research on QDSSC can be further extended by modifying and optimizing each component. Some of the suggestions for future research are as follows.

• Different novel QD sensitizers can be developed for enhancing light absorption by co-sensitization approach with minimal spectral overlap.

• Stability of QDSSC can be enhanced by replacing liquid electrolyte with solid electrolytes or quasi-solid electrolytes.

• Replacing FTO substrate with highly flexible substrate for flexible QDSSC.

• QDSSC performance can be further improved by developing more hierarchical QD based tandem solar cell.

• CEs with improved electrocatalytic ability are an area to be studied that can be a breakthrough in enhancing efficiency of QDSSCs.

• As a strategy to improve stability of QDs in QDSSC, QDs with surface coating can be further investigated as long term stability of QDs are highly important in realizing the commercialization of QDSSCs.